DARWIN'S DICE

Darwin's Dice

THE IDEA OF CHANCE IN THE THOUGHT OF CHARLES DARWIN

Curtis Johnson

Oxford University Press is a department of the University of Oxford.
It furthers the University's objective of excellence in research, scholarship,
and education by publishing worldwide.

Oxford New York
Auckland Cape Town Dar es Salaam Hong Kong Karachi
Kuala Lumpur Madrid Melbourne Mexico City Nairobi
New Delhi Shanghai Taipei Toronto

With offices in
Argentina Austria Brazil Chile Czech Republic France Greece
Guatemala Hungary Italy Japan Poland Portugal Singapore
South Korea Switzerland Thailand Turkey Ukraine Vietnam

Published in the United States of America by
Oxford University Press
198 Madison Avenue, New York, NY 10016

Library of Congress Cataloging-in-Publication Data
CIP data is on file at the Library of Congress
ISBN 978–0–19–936141–0

9 8 7 6 5 4 3 2 1
Printed in the United States of America
on acid-free paper

For my family

Contents

Acknowledgments

THE SYSTEMATIC STUDY of chance in nature's operations has an ancient pedigree, dating back at least to the writings of Empedocles (5th c. BCE) and Aristotle (4th c. BCE). If one were permitted to acknowledge debts to the no longer living, I would begin by acknowledging them. It is not simply that they wrestled with the idea of chance in nature; they did so in just the way Darwin did. Darwin himself was not directly influenced by these authors. I doubt he ever read anything by Empedocles, and his acquaintance with Aristotle's writings came only late in his life, long after he had developed and refined his own ideas. But these two ancient thinkers put a Darwinian sort of chance into a long tradition of thought that is with us still.

Any work of this sort could not be undertaken without the insights of other scholars laboring in the fields. It would be impossible to give an exhaustive account of what I owe to all of these men and women, but I refer the interested reader to the bibliography for a register of most of my debts to other authors. Several of these stand out for special mention: the volume of collected essays edited by Gigerenzer, G. et al., eds., 1989; a two volume work edited by Krueger and Morgan, 1987; and several individual works by those whom I regard as leading authorities in the study of chance in evolutionary biology today, especially John Beatty, James Lennox, and Roberta Millstein. I have no personal acquaintance with any of these authors but my work has benefited beyond measure by the foundations laid by them.

I take special pleasure in acknowledging Darwin scholars who have read some or all of the ms. and made invaluable suggestions, all of which have led to improvements. I am especially grateful to my father, Dr. Albert Johnson, an expert in ecology, who read the entire ms. and provided keen advice at every stage; J. David Archibald, who, with a firm touch, took the training wheels off the bike; Michael Ghiselin, who from the beginning has been unstinting in his encouragement and willingness to share his knowledge of all things Darwinian; and David Depew, whose comments on an earlier draft of the central ideas contained herein awakened me to subtleties I had not yet seen. My debts to all four run deep. Any remaining defects are attributable to me.

I give special thanks to two of my former undergraduate students at Lewis & Clark College, Julia Eckes and Kris Lyon, both of whom helped immeasurably with editing and editorial assistance at various points in the preparation of the manuscript.

I also must convey my gratitude to the editorial staff at Oxford University Press: Hallie Stebbins, who was the first to take notice of and interest in the project; Jeremy Lewis, who seamlessly took over where Hallie left off; Erik Hane, my day-to-day contact at Oxford, and to Jayanthi Bhaskar for her grace and professionalism in shepherding the work through the labyrinthine details of the production process. I also thank the several external reviewers for Oxford U.P. who read and commented on the ms. with erudition and sensitivity.

My greatest debts are to my wife and daughters Sophia and Alexis, and my parents for constant encouragement and support. Sophia is the artist who produced the illustrations for the chapter headings; Alexis created the graphic on p. 59. Above all, I could not have undertaken this work without the untiring support and assistance in every manner of way—too many to enumerate—of my wife Loretta.

Introduction

CHARLES DARWIN'S "BIG idea" is generally thought to be his discovery of the mechanism of natural selection in evolution. That discovery was without question a big idea. But, as Darwin himself often confessed, natural selection cannot work without prior variations in the organisms that will be selected or not for survival. Whence the variations, or at least what did Darwin believe about this? That is the question I examine in what follows. Darwin thought "variations" are in many or most cases "just by chance." I hope to show what he meant by this expression and what he believed the implications are if one accepts it. "Chance variation" may have been an even bigger idea for Darwin than natural selection, or so I shall attempt to show.

Thus, whatever "Darwinism" is, this is not a book about Darwinism. Nor is it a book about contemporary evolutionary theory or the "new synthesis" or the "extended synthesis."[1] It is rather a book about "chance" in Darwin's writing. To that extent it must confront "Darwinism" more broadly, even in its recent and contemporary incarnations, if only to situate the problems it deals with in a proper context.

But an answer to "what is Darwinism" is surprisingly elusive. Even if we grant that "Darwinism is whatever Darwin said it is," the problems of identifying Darwinism only begin there. Darwin's views seem to have evolved over the course of a long and prolific career as a scientist, author, and correspondent. Did he maintain a consistent position? Was "Darwinism" always the same thing to Darwin? Many scholars who have considered this question, perhaps most, have held and defended the view that his ideas did change, so that "Darwinism" as "whatever Darwin said it is" is a moving target (Lennox, 2004; Beatty, in Mueller and Pigliucci, eds., 2010; Hull, in Kohn, ed., 1985; Ospovat, 1981; Browne, 1982; Gayon, 2003).

To mention only the most notorious example, the *Origin of Species*, from its first appearance in 1859 through the last edition (sixth, 1872) underwent revisions, many of decided significance, for each new edition (cf. *Origin* 1959, *Variorum* edition, editor's introduction; and for additional discussions, Moore, 1979, 133; Vorzimmer, 1970, *passim*; Ruse, 1979, 210–11; Browne, 1995, 2002, 2006).[2] But that is only the beginning of the story. From the time of his *Notebooks*, containing Darwin's earliest reflections on the origin of species question (1836–1842), through his late book on earthworms (1881) Darwin tinkered almost obsessively with the way he chose to present his theory. This fact raises the suspicion that he did in fact change his mind and that he came to make "adjustments" to the theory along the way.[3]

My view is that "Darwinism" had a single meaning to Darwin from beginning to end. Yes, changes were made in exposition, over and over again, and in one sense, as a philosophical platitude, one cannot change one's way of saying something without changing what one says, and therefore what one is taken to mean.[4] But another way of thinking about changes in presentation is to ask whether the author intended a change in meaning, especially as regards core beliefs and elements of a theory. In this regard, according to the well-established "principle of charity" in philosophical studies, one looks for "philosophical consistency" and decides that a significant shift in outlook has occurred only when forced to do so by irrefutable evidence.[5]

In the case of Darwin such evidence is hard to come by. Even explicit statements by Darwin such as "I probably underestimated the importance of factor 'x' (e.g., so-called 'use-inheritance') in earlier versions of my theory," do not rise to the level of strong evidence for a shift in basic outlook, yet that is the sort of evidence upon which arguments for a shift is often based.[6] Placed alongside a considerable body of contrary evidence, the case for an ever-changing "Darwinism" is weakened. One is not free to ignore the evidence supporting a Darwinian change. But one may overcome it if counter-evidence is available and if a better explanation for the supporting evidence is available.

On the other hand, Darwin did change his mode of exposition, repeatedly. An examination of the Darwinian corpus shows that many of the most important changes centered on how he wished to present the role of "chance" in evolution to an ever-expanding reading public, especially after the *Origin* first appeared. But the changes started to appear much earlier, from unguarded reflections in the 1836–1839 *Notebooks* to the more publicly attuned 1842 "Sketch" and 1844 "Essay" (Kohn and Stauffer, 1982) all the way through the several editions of the *Origin* and beyond. Something deliberate is going on here, and to discover what that something is motivates the work presented here. To anticipate, I wish to make and defend the following claims:

(1) Darwin discovered "chance" as a basic factor in evolution from an early time in his career, perhaps mid-1837.[7]
(2) Darwin understood some important implications of this discovery from a nearly equal early period for how his views would be received, specifically: (1) that "chance" (in its primary meaning for Darwin) would be regarded as a "dangerous" idea (in this he was correct); (2) that he probably had to readjust his own religious views in light of his discovery; (3) that he could not in good conscience pretend to himself or the world that he did not really mean it; (4) that to ensure scientific acceptance of his discovery he would need to cast the role of chance in ways that, while preserving its central meaning, would either obscure its role in his theory or at least make it seem innocuous to otherwise friendly natural philosophers and scientists; and (5) that to accomplish this end he would need to rework his wording in his published writings.
(3) Changes made by Darwin in how he chose to present "chance" in his theory may be of greater significance than any others in the Darwinian corpus. At a minimum they are extremely important in seeing how he "evolved" in mode of expression.

I understand that these are strong and for some readers controversial claims. Whether they can be supported will depend on the evidence from Darwin's oeuvre. I present what evidence I could assemble in the following pages and must let the reader decide whether the evidence and supporting argumentation are up to the challenge.

It may fairly be asked, against whom am I making these claims? Darwin's thoughts, as many commentators have observed, are often refracted through the lens of the reception of his views as they became more widely known by a broader reading public. For example, M. J. Hodge has suggested that Darwin's "reception" among a large reading public was created and promoted by his son Francis in published letters edited by Francis, as well as in Francis's own interpretive essays—the so-called Franciscan interpretation. Francis did not alter Darwin's words but he did make choices about what should be published after his father's death and what should remain private. Later commentators have offered different interpretations of what Darwinism meant to Darwin as more of Darwin's "private" writings became public. The point here is only that how a thinker is understood is often a result as much about what others have said as it is about what the thinker wrote himself.

I have not been motivated in this work by a desire to show disagreements with previous interpretations, although, inevitably, my arguments will challenge many of them. But I got my start in Darwinian studies by examining only the words and works of Charles Darwin. I thus started out with no axes to grind. I believe this is an advantage, insofar as my readings were not at first influenced by the readings of other scholars. Naturally, over the course of years of engagement with the vast literature on Darwin and Darwinism I have come to see where my work intersects, overlaps, and disagrees with this literature. I try to give a full account of my many debts to the scholars who have been laboring in the field and to show where my arguments differ from theirs. But I try to set out and move through with a clean slate, basing my claims on Darwin's *ipsissima verba* rather than on what others have said.

DARWINISM

The Darwinian theory, usually rendered in shorthand as the theory of "descent with modification by means of natural selection," may be reduced to a syllogistic core that goes something like this:[8]

(1) Variation. All creatures that reproduce (sexually or asexually—it doesn't matter[9]) will produce offspring that *vary* slightly from themselves. An offspring might have slightly longer legs, or a slightly shorter beak, or slightly more hair, than its parents, and so it is said to vary. It is important to say that Darwin often claimed that he did not know *how* or *why* variations occur, only that they do occur. No parents' child is identical to its parents, but how it will vary no one can predict. Darwin could often do no better than to say that any variation from parent to child is due to what we must, in our ignorance, call *chance*.

(2) Heritability. Variations are often passed along in reproduction. Children with longer legs or more hair are likely to have children with these same traits, or even with these same traits more pronounced, and so on down the line of generation. In other words, variations often have a tendency to be *preserved*.

(3) Competition for survival. More creatures are born in every species or group than can normally survive. They reproduce faster than the resources upon which they depend for sustenance. Therefore, some—actually many—must perish, as a regular fact of life. Only the few ever survive. This phenomenon came to be called by Darwin "survival of the fittest," an expression that was invented by Herbert Spencer and brought to Darwin's notice by the co-discoverer of the theory of natural selection, A. R. Wallace.

If these three events occur in nature—and Darwin was certain they did—then the mechanism of Natural Selection would allow evolution to happen.

(4) Natural Selection. This principle determines who are the winners and losers in the perpetual struggle for existence. Those creatures that have varied in "favorable" directions are more likely to survive than those that have not varied, or have varied in unfavorable ways. For example, in a climate where longer hair provides a better protection against death by freezing than shorter hair, a variant individual with longer hair will be "selected" by nature to survive against its rivals who have been born with shorter hair, and this successful variant is likely to pass along the winning trait to its offspring. Again, Darwin did not claim to know how or why some individuals happened to vary—happened to be born with longer hair in our example—but only that if they did vary in favorable directions, they had a better chance to be selected for survival than ones that did not vary.[10]

Of these four parts or ideas of the theory, this book is mainly about the first—variation—and even more narrowly, only variations that Darwin attributed to chance. The other ideas, of course, are fundamental to the theory, and no one believes that Darwin ever wavered from his belief in them, or in the primacy of natural selection among other factors that play a role in evolution. What is usually at issue in arguments for a changing Darwinism, rather, is the role played by "chance" in explaining variation. This idea more than any other sets Darwin's theory apart from all other evolutionary theories in his day, and thus is important for establishing Darwin's theory as distinctively "Darwinian." The idea of "chance," and the role it plays in the modification and "transmutation" of species, remained steadfast and the same in Darwin's thought from his first revelations in 1837–1838 about what goes on in nature to all subsequent works where he addressed the question.[11] It is also, as I shall show, the one part of his theory that underwent the most dramatic changes in exposition.

These changes, directly and indirectly, account in turn for most of the suspicion that Darwin actually changed his mind, even though those who bring forward this argument have not been entirely clear about the importance of this shift for their own arguments. For example, one typical argument is that Darwin became more "Lamarckian" over the years.[12] This is generally taken to mean that he came to strengthen a role for so-called "use-inheritance" in evolutionary change. What generally goes unnoticed in these accounts is that "use-inheritance" can only be strengthened by diminishing a role for something else, and that something else is usually "chance." In fact the impression that Darwin strengthened "use-inheritance" is generated in part by the fact that he did (in words) reduce or even disguise the role that he had earlier assigned to chance. But if he did not really change his mind about chance, he did not really change his mind about use-inheritance.

DARWIN'S APPROPRIATION OF CHANCE

Chance as an important factor in how to understand nature was not Darwin's unique discovery. Philosophers and naturalists had much to say about "chance," even in quasi-evolutionary contexts, long before Darwin, as is the case, for example, with the Greek philosopher Empedocles (4th c. BCE), as recorded in Aristotle's *Physics*. Aristotle too considered what sort of role chance might be said to play in natural events.[13] But the idea that chance might play a role in shaping the organic world, in such a way that random variations paradoxically give rise to apparent design and order, was no part of the scientific mainstream of Darwin's day. If anything, chance was anathema to most scientists and philosophers (cf. D. Hull, 1973, 15, 55–68; Browne, 2006, 92–3). To most thinkers, chance connoted a variety of ideas that seemed contrary not only to revealed and natural religion but also to common sense. Even skeptical thinkers like David Hume, who made serious efforts to consider the possibility that chance may have some role to play in nature, came to reject it (*Dialogues Concerning Natural Religion*, 1779, Pts. II, IV, V, VII, XII; cf. D. Dennett, 1995, 28–33; Dawkins, 2006, chapter 4). Most people did not even consider the possibility. The many writers who most influenced Darwin's thinking about nature—men like Lyell,

Whewell, Herschel, and before them William Paley, with their deep admiration for nature's orderliness and evident design—dismissed a role for chance out of hand.[14]

Darwin, by contrast, understood from his earliest reflections on the origin of species question in 1837–1838 that he would be required by the tenets of his science to make room for a role for "chance" in the evolution of new species. Our question concerns how he handled this issue. "Chance" as Darwin used it was a bogey for most of his audience, friendly and unfriendly alike. Chance, at least in one important sense, means fortuity, and most people in Darwin's day, and even now, could not accept a world in which fortuity played a guiding role in evolution.[15] Yet Darwin believed fortuity was at the very core of modifications leading to the origin of new species.[16] The implications of any such view were significant. The earth, its geological features, and its organic inhabitants are here only through lucky accidents? For many people that was a hard pill to swallow. Darwin did accept it, but also knew he would have to get his audience to accept it too if he were to succeed in establishing his theory as the correct account of the origin of species.

Darwin realized he would need to tread carefully. His early public presentations of the theory, especially in the *Origin* itself, were not careful enough. Under the onslaught of criticism that the *Origin* received after its first appearance in 1859, Darwin decided that he needed to downplay, or perhaps better disguise the role of, chance if his theory were to be generally accepted.[17] In light of this recognition he adopted a variety of rhetorical strategies that added up to a deliberate campaign to retain chance as a central element while making it appear to most readers that he did not; or, as with the "stone-house" metaphor (discussed in chapter 7 of the present work), making it appear less "dangerous" an idea than many supposed.

DARWIN'S RELIGIOUS VIEWS

Darwin's early recognition of chance in causing variation also has implications for how we understand the evolution of his religious beliefs. The customary view, based mainly on his *Autobiography* and the small selection of letters that was available to a large audience previous to the mammoth "Correspondence Project" (1985 to present), is that he gradually shifted from "early orthodoxy" to a "liberal form of theism," and then in later years "into an agnosticism tending at times toward atheism" (Herbert, 1974, 232; Moore, 1979, 314–15; Ruse, 1979, 180–4; Ruse, 2010, 1–8; Lennox, 2004; R. J. Richards, 1989, 77–7; N. Gillespie, 1979, ch. 8; Beatty, 2006).[18]

It seems probable that his departure from Christian faith was earlier, more abrupt, and more complete than this view indicates. The reason for thinking so stems from the same source that so many of Darwin's contemporaries rejected a role for chance in nature's workings: a chance-governed world seems tantamount to a godless world. Einstein made this very connection himself 75 years later when he famously said, "God does not play dice with the universe."[19] Darwin undoubtedly understood this implication of his theory, but rather than conclude that chance plays no role in nature he appears to have concluded instead that God does not have much to do with nature at all.[20]

How new this idea was in Darwin's day is suggested by a quote from one who could not accept it. Charles Kingsley (a distinguished professor of History at Cambridge University and correspondent with Darwin), no doubt reflecting a common view, observed in 1871, "God is great, or else there is no God at all" (in Moore, 1979, 339; for Moore's analysis of Darwin's religious views in the *Notebooks*, 319–25 and nn. 56–87). The 1871 comment of St. G. Mivart's (a younger aspiring biologist and devout Catholic) was more pointed: "Unhappily the acceptance of your views means with many the abandonment of the belief in God and the immortality of the soul" (*CCD*, vol. 19, 36). Unlike Mivart, Kingsley, and many others, Darwin appears to have adopted the second half of the disjunct: not that "God is great," but rather "there is no God at all."[21]

Some students of Darwin's thought will wish to make an objection to this claim. But let us look at Darwin's words. In May 1860, only a few months after the *Origin* first appeared, Darwin had this to say to his early American supporter, the Harvard botanist Asa Gray:

> With respect to the theological view of the question; this is always painful to me.—I am bewildered.—I had no intention to write atheistically. But I own that I cannot see, as plainly as others do, & as I shd. wish to do, evidence of design and beneficence on all sides of us. There seems to me too much misery in the world. I cannot persuade myself that a beneficent & omnipotent God would have designedly created the Ichneumonidae with the express intention of their feeding within the living bodies of caterpillars, or that a cat should play with mice. Not believing this, I see no necessity in the belief that the eye was expressly designed. On the other hand, I cannot be contented to view this wonderful universe, and especially the nature of man, & to conclude that everything is the result of brute force. I am inclined to look at everything as resulting from designed laws, with the details, whether good or bad, left to the working out of what we may call chance. Not that this notion at all satisfies me. I feel most deeply that the whole subject is too profound for the human intellect. A dog might as well speculate on the mind of Newton.—Let each man hope and believe what he can.

Darwin continues in the same letter:

> Certainly I agree with you that my views are not at all necessarily atheistical. The lightning kills a man, whether a good one or a bad one, owing to the excessively complex action of natural laws.—A child (who may turn out an idiot) is born by action of even more complex laws,—and I can see no reason, why a man, or other animal, may not have been aboriginally produced by other laws; & that all these laws may have been expressly designed by an omniscient Creator, who foresaw every future event & consequence. But the more I think the more bewildered I become; as indeed I have probably shown by this letter (*CCD* 8, 223 [May 22, 1860]).

The question for Darwin came down to whether the notion of an "undesigned nature" made any sense. Two months later, after continuing to ponder, Darwin wrote again to Gray, and again showed that he was still in a quandary:

> One more word on "designed laws" & "undesigned results." I see a bird which I want for food, take my gun and kill it, I do this *designedly*.—An innocent & good man stands under a tree and is killed by a flash of lightning. Do you believe (& I really shd. like to hear) that God *designedly* killed this man? Many or most persons do believe this; I can't and don't.—If you believe so, do you believe that when a swallow snaps up a gnat that God designed that that particular sparrow shd. snap up that particular gnat at that particular instant? I believe that the man and the gnat are in the same predicament.—If the death of neither man nor gnat are [*sic*] designed, I see no reason to believe that their *first* birth or production shd. be necessarily designed. Yet, as I said before, I cannot persuade myself that electricity acts, that the tree grows, that man aspires to the loftiest conceptions all from blind, brute force (*CCD* vol. 8, 275 [July 3, 1860]).

Well, which is it, designed laws, or "blind, brute force," with no foresight or intention by any designer about what laws should be created or how they would operate? For anyone who wonders what Darwin really believed, the passages just quoted give no clear answer. "Brute force," which Darwin could not fully accept as an explanation for life's diversity, apparently refers to the means by which variations arise, and may equally well be rendered by the word "chance." He had no doubt that once variations did arise they would be pruned and preserved by non-intelligent processes. But how did they arise in the first place?

Darwin continued to express his doubts and uncertainties to other correspondents in the months after the *Origin* first appeared. For example, to Charles Lyell in

August 1861 Darwin responded with skepticism to Gray's argument that the course of streams are "designed" by an intelligent maker:

> I doubt whether I have made what I think clear; but certainly A. Gray's notion of the course of variation having been led, like a stream of water by Gravity, seems to me to smash the whole affair. It reminds me of a Spaniard whom I told I was trying to make out how the Cordillera were formed; & he answered me that it was useless for "God made them." . . . I must think that such views of Asa Gray & Herschel merely show that the subject in their minds is in Comte's theological stage of science[22] (*CCD* vol. 9 [August 1, 1861]).

This was a bold claim for Darwin to make to his mentor and friend Lyell, whose own religious convictions that God made the laws governing geological nature were often affirmed in his great work *Principles of Geology* (1830–1833). Darwin must have known that he was challenging Lyell's deepest theological convictions. Perhaps to soften the blow, Darwin added the following:

> The view that each variation has been providentially arranged seems to me to make natural selection entirely superfluous, & indeed takes the whole case of appearance of new species out of the range of science.—It seems to me that variations in the wild and domestic conditions are due to unknown causes & are without purpose & insofar accidental; & that they become purposeful only when they are selected by man for his pleasure, or by what we call natural selection in the struggle for life under changing conditions. I do not wish to say that God did not foresee everything which would ensue; but here comes very nearly the same sort of wretched embroglio as between free-will & preordained necessity (*CCD* vol. 9 [August 1, 1861]).

These letters to his close associates must be taken to disclose some of Darwin's deepest thoughts in 1860–1861 on the subject of divine intelligence in the creation of species. What do the passages tell us? Strictly speaking they tell us three things that are not mutually consistent: (1) that Darwin had never *intended* to write "atheistically" (by itself, of course, that does not mean he had no atheistical leanings); (2) that he can see no evidence of "design" (or, therefore, an omniscient God) in natural productions; and (3) the whole subject of design and God is too profound for him to know what to believe. Perhaps God created natural laws at the beginning and then left the world alone so as to allow His laws to play out according to an invisible divine plan. But it is not clear that Darwin really believed that. His ultimate refuge in all the passages reproduced above (and many more) was that the whole question was too profound for him to take it on (see chapter 3 below for a further development of this subject).

MEANINGS OF CHANCE

"Chance" in evolutionary biology is not just one among a wide range of concepts that requires systematic attention. It is, in one of its several meanings, an entire way of thinking about nature, one that in this case separates Darwin from most other systematic biologists of his own time and most of those prior to him. It also separated him from a number of naturalists who came after him and were especially offended or disturbed by his views about the place of chance in his theory.

The role of chance in the biological sciences was not Darwin's original brainchild; it had been studied in biological contexts since Greek antiquity. But Darwin, more than anyone else, brought it into nineteenth-century biological theory, and it remains a subject of interest and importance in evolutionary theory today (see chapter 1). Understanding Darwin's evolving views and expressions about chance helps illuminate some modern controversies. It also shows how one person—Darwin—confronting the dilemmas posed by his discovery of chance, negotiated the difficulties, and emerged at the end of his career with formulations that were strikingly different from those with which he began. Eventually, by the time of the publication of his *Descent of Man* (1871) and *Expression of Emotions* (1872), he simply stopped talking about chance altogether. The word and the idea, no matter how expressed, had evidently become a quagmire he could do without.

TWO MEANINGS OF CHANCE IN DARWIN'S THOUGHT

Chance is a complex concept that has various applications in human thought and speech.[23] It can mean anything from "coincidences" to "probabilities" to "propensities" to "degrees of rational belief," and more.[24] In Darwin's writing it tends overwhelmingly to mean one of two things: the *likelihood* or *probability* that some organisms will survive in the struggle for existence (this is related to their *adaptedness* to existing conditions); and the *fortuity* that new variations will be well adapted to their conditions, or better adapted than their rivals (Hodge, in Krueger, 1987, vol. 2, chapter 10; Lennox, 2004). The former idea may be rendered in statistical terms, or, if one prefers, as a percentage. For example, an evolutionary biologist might cast the likelihood of a well-adapted variant to be 90% greater to survive and reproduce than a poorly adapted one (e.g., Brandon, 1990, 11; Lennox, 2004). Empirical confirmation is available, at least in principle, even if the matter is complicated. The latter idea is better characterized as a stochastic concept: what variations occur in nature just happen to occur, with no probabilities being assigned either for their appearance or non-appearance. The statistical idea is subject to precise measurement (again, in principle, as a probability), the latter is not.[25] Darwin himself did not employ rigorous statistical analyses of his "chances of survival" idea, although he was on the track of doing so when, for example, he sent questionnaires to field biologists asking about their observations of natural processes. But he did not even begin to apply probability theory to the causes of variations. He made guesses about what these causes are but he unfailingly admitted that an understanding of such causes could often be no more than guesses.

"Probability of survival" and "fortuity of variation" are connected concepts (because both refer to and depend upon the idea of adaptation), but they are fundamentally different. I explore the distinction more fully in chapter 1. But as a preliminary, I draw attention to the following concerns. First, much of the recent scholarship on "chance" as a concept in scientific studies centers on statistics and probability theory (Gigerenzer et al., 1989, *passim*; Krueger et al., eds., volume 2, 1987, chapters 10–13; Millstein, 2011, 425 ff.; Lennox, 2004). Darwin grasped that some organisms, because of their structures and habits, have a greater probability of surviving than others. This is an important claim, no doubt. But Darwin did not employ a statistical analysis of this phenomenon, or even pretend to. It seems that he was not well acquainted with probability theory or statistics, and it is evident that he did not pretend to study his claim with scientific rigor. He simply ventured the opinion that some organisms have a better chance of survival than others because it was a necessary deduction from other aspects of his theory. If variations occur, if a struggle for survival is a part of natural history, if the more fit tend to survive against the less fit, then it must follow as a matter of logic (even without empirical confirmation) that the "chances" for survival will favor the more fit. Darwin's claim has stood up to empirical scrutiny over the 150 years since he wrote by botanists, zoologists, and other natural scientists, but Darwin himself did not do the mathematical and statistical analyses that would be required today for empirical verification.

A second concern is whether "cause unknown" (for any variation that appears) can really be called "chance" or can be a source of concern to Darwin's intended audience. The issue may be divided into two parts. The first is whether "cause unknown" really means, or to Darwin did mean, "chance cause." The other is whether the idea of an "unknown cause" was as innocuous an idea as it might first seem to be.

As to the first, some people might dispute that "cause unknown" is the same as "chance" because the former refers only to a lack of knowledge on the part of observers studying the phenomenon, whereas the latter seems to suggest "no cause." Darwin sometimes suggested that he meant nothing more than "our ignorance" when he said variations come about "by chance" (*Variorum*, V. 4–5). I do not wish to argue that Darwin was being dishonest or disingenuous in this suggestion. He was committed to the idea that all natural events (including variations) have "causes," and thus are fortuitous only from the standpoint of human understanding. No alarm here, even for a theologically disposed audience.

But "chance as unknown cause" is not everything or even mainly what Darwin believed. The expression "unknown cause" suggests a cause that *can* be known, even if at present the cause is not known. If the cause *were* known, and it was a theologically acceptable cause, for example a directing agency such as God, Darwin's sympathetic readers would have had nothing to fear about the theory. But Darwin's thought was more radical. By "unknown cause" he implied in his more private and less guarded moments that the cause of at least some variations is *unknowable*, even in principle (Beatty, 2006, develops the distinction). And the reason for the unknowability of such causes is not lack of human understanding but rather a lack of a directing rational agency. When Darwin looked at nature's productions, especially the appearance of new variations, he could not fathom any conceivable reason in many cases for why those that occurred did occur or why some that might have been expected to occur, on any customary notion of cause, did not occur. Variations were often, as far as Darwin could tell, simply random with respect to survival needs: some would survive, some would perish, but variations could not be "designed" or be considered "rational" with either of these outcomes in view (cf. *CCD* vol. 9 [August 1, 1861], 225, letter to Lyell).

Darwin's understanding of chance in this latter sense (random with respect to future adaptive needs) did much to separate Darwin's thought from the tradition of natural philosophy he had learned from his mentors. Causes "make sense" from the standpoint of what can be expected in nature and of the ability, therefore, to make accurate predictions. No doubt Darwin believed the natural order "made sense," insofar as it "holds together." But it did not make sense in terms of how it held together from "causes" as normally understood. If variations are just thrown up at random, like the way dice fall in a given toss or cards turn up in a blind draw, nature's order becomes a major puzzle.[26] Random variations would seem to lead to the idea of an unordered or even chaotic world. Darwin did not believe the world was chaotic. Indeed, he always admired the order he found in it. But he could not believe the world was governed by a presiding rational intelligence that ensured its order. He was thus forced to confront one of the major dilemmas of his life: how can one get order out of random processes? His solution was, of course, the idea of natural selection. But the solution would only be a solution if the prior step in transmutation, variation, was itself random. Natural selection is a breakthrough idea only against the background of "chance variations." One may abbreviate by saying Darwin's theory gets its critical bite and novelty by its ability to explain how order comes out of fortuity without invoking "higher powers."

Darwin, however, never did abandon the idea that variations are "caused." He was simply at a loss about how to give a comprehensive account of those causes. He did try to say what they are, and offered valuable suggestions (the subject of chapter 3 in the present work). But often he was reduced to puzzlement. Understanding has naturally progressed since Darwin wrote. The rediscovery of Mendel's writing in the early part of the twentieth century, and the many advances in mathematical, cellular, molecular, and genetic biology since then, have improved biologists' understanding of what kind of question Darwin was up against in his search for causes and where to look for them. But even today it is impossible to get away from the language of "chance" in variation. The *mechanisms* of variation are better understood than ever, but the ability to predict what variations will occur and what will not is not much better off than when Darwin wrote. So we still encounter in the leading works in evolutionary biology today the ideas of "random drift," "sampling error," "fortuitous recombination," "point mutations," and the like. Many scholars continue to insist that biological evolution—what species will someday appear and what ones will not—is mostly guesswork. One finds few biologists today attributing variations to a directing intelligence. They still prefer to say "chance."

If Darwin's ideas of "chance" caused worry and alarm in his reading public, it would have been chance only in the sense described here—"randomness with respect to future adaptive needs." Unlike "unknown causes" that might nevertheless be known someday, Darwin's chance seemed to his reading audience to rule out "knowability," and this in turn ruled out intelligence and design. That idea did frighten many people, as I show below. Darwin could not duck the controversy by either denying chance or allowing intelligent design. He was stuck with "chance as randomness" whether he was happy with it or not. The present work focuses on chance in this restricted sense. This is the point of controversy and the element of his theory that must be credited with making his idea a "dangerous" one to many of his readers.

WORRIES ABOUT CHANCE

Why was Darwin's theory of "evolution through natural selection" successful as the correct account of how biological diversity arose? After all, Darwin was by no means the first evolutionary thinker. Even "natural selection" was not altogether new with him, as he candidly acknowledged (a bit after the fact) in his "Historical Sketch," appended to the first American and third English edition of the *Origin* (both in 1861), retained and enlarged in subsequent editions (Johnson, 2007). What truly was novel in Darwin's approach was making room for "chance" in the modification of species and their transmutation into new species. The three most well-known evolutionists prior to Darwin—his grandfather Erasmus, the French zoologist J. B. Lamarck, and the anonymous author of the sensational *Vestiges of the Natural History of Creation* (Robert Chambers)—all produced accounts of evolution that not only did not require chance but in fact ruled out a role for it. Evolution of new species, they believed, was guided and directed, either by a superior intelligence or by innate faculties and potencies within living organisms that enabled them to adapt to changing conditions.[27]

If anything distinguishes Darwin's theory from these earlier versions, it was his focus on variation.[28] What Darwin discovered is that variations appear to occur "in all directions," not just favorable, let alone progressive ones (e.g., *CCD* vol. 8, 340 [September 1, 1860], to Lyell; ibid., 342 [September 2, 1860] to Hooker; ibid., 355 [September 12, 1860] to Lyell). From this perspective, what variations occur is "just a matter of luck," good, bad, or indifferent. Above all, it is impossible to predict what variations will happen to come along, and the ability to make predictions, Darwin knew, is a hallmark of a good science (Darwin, 1958, 109; and Hull, 1973, 32–3).

But, even granting that variations are "caused," to *say* that they are "by chance," as Darwin often did, caused alarm.[29] Even before the *Origin* first appeared or before anyone had any idea about Darwin's theory, philosophers, scientists, and theologians were worried about any possible role for chance in the organic or inorganic world (Ruse, 1979, 71–4; also Moore, 1979, 106–10—citing Lyell's "disquiet and alarm," and similar concerns voiced by Romanes, Butler, and James, going back to 1855, cited in J. Moore, 1979; Gigerenzer et al., 1989, surveys the "empire of chance" from the seventeenth to the twentieth centuries in a variety of scientific disciplines). What was the worry all about? If it needs to be spelled out, the worry was the implications of this view for a designed world, and by extension, for an omniscient deity who presided over the design.

Several strands of pre-Darwinian thought, whatever the provenance, converged to make the concern of an undesigned universe a centerpiece of criticisms of chance. Whether species had evolved or not—and differences of opinion on this point were rife even before, not to mention after the *Origin*—almost no one could accept chance as a factor in the process.

The worries about chance, however, grew decidedly more pronounced after the appearance of the *Origin*, thanks in large part, no doubt, to Darwin's employment of the word to describe how variations at least sometimes come about.[30] The first edition of the *Origin*, for example, invokes "chance" as a "cause" of variation at least a dozen times—even though usually Darwin took pains to qualify the notion so as to make it mean, in effect, "cause unknown," and simultaneously to make clear his opinion that "unknown cause" does not mean "no cause."[31] But concerns persisted. This point was forcefully brought home to Darwin in 1871, in a letter from the cleric and botanists George Henslow (son of Darwin's friend John Stevens Henslow). Henslow implored Darwin to condemn in writing the "wild and false assertion" made by many of his clerical brethren and other evolutionists that Darwin was "an *Atheist* and all the rest of it!" (*CCD* vol. 19, 713, original emphasis).

> I should very much like to hear from you [Henslow continued] if the following impression which I have is correct: viz., that when you speak of *Chance* in connection with Nat. Selection you leave it to be *understood* that *higher natural* laws (but undiscoverable at present) cause the issue of specific forms out of those "chance" variations (ibid., original emphasis).

Darwin politely declined to assuage Henslow's concerns, and instead referred Henslow to his earlier public statements on the matter (ibid., 714 and n. 2). Those earlier published statements, as we shall see, neither affirm nor deny the charge of atheism. Rather they say simply that Darwin finds the question too profound for his limited intellect to answer (e.g., *Variation* vol. 2, 431–2).

Henslow's worries were just those of other scientists friendly to Darwin, such as Charles Lyell and Asa Gray, who accepted a Darwinian account of evolution by means of natural selection. It was not evolution to which they objected, nor even the mechanism of natural selection, but the prior step—"chance" variations (Gigerenzer et al., 1989, 132–6). This idea would not do. The variations, they believed, must be "guided" or designed. The stumbling block was unguided variations. If that idea could be replaced with variations directed to be what they are by a higher power, the rest of the theory would get a clean pass (cf. Lyell, 1863, 506). Darwin decided to ride out the storm of criticism rather than capitulate.

NOTES

1. For readers interested in current developments in evolutionary theory, a good place to begin is M. Pigliucci and Gerd Mueller, eds., *The Extended Synthesis*, MIT Press, 2010.

2. All references to Darwin's *Origin of Species* are to the *Variorum* edition published in 1959 under the editorship of Morse Peckham. The references are to the *Variorum* page number, followed by the chapter number in Darwin's text (given in roman numerals), followed by the line number of the passage as given in Peckham. Small italicized letters *a, b, c, d, e,* and *f* at the end of Peckham's line numbers refer to the number of the edition of the *Origin: a* as the first edition is not used*; thereafter, b* = second, *c* = third; and so forth.

3. Herbert (1974, 226); Moore (1979, 127, 145, 176); Vorzimmer (1972); E. Mayr (1982, 690–3); E. Mayr (1963, xxiv-xxvii); P. Bowler (1983, 67); M. Ruse (1979, 211); P. Bowler (1993, 49); J. Browne (2003, 208, 283–4, 315, 354, 369, 407); *Correspondence of Charles Darwin* (hereafter *CCD*), vol. 11 (1999, 137 n. 6); J. Costa (2009, 494–5); Beatty (2010, 22). H. E. Gruber (in Kohn,

ed., 1985, 17) refers to "the ambiguities in Darwin's position [about the species question] at *every* point in his development."

4. Darwin of course did modify his theory as he went along. It did not sprout full-grown from his mind overnight. The "complete theory," whatever that might mean, was years in development. This aspect of Darwin's thought has received a great deal of scholarly attention, especially in the last 25 years as more of his private notes and correspondence has been made available to a broad reading public (e.g., several essays in Kohn, ed., 1985, including especially chapters 1–9 and chapter 26, and detailed references within those chapters; cf. also Kohn, 1980; Kohn, Smith, and Stauffer, 1982; F. Sulloway (1982); S. Herbert, 1974, 1977, and references; Sloan, 1986). My assertion pertains to the "syllogistic outline" of the theory, and more particularly to Darwin's views on the role of chance in variation.

5. I shall not debate in this work the question of whether it is possible to assign "authorial intention" from an examination of an author's oeuvre. I shall accept as a premise that it is possible to make plausible inferences about what Darwin intended, granting that the premise is sometimes contested. For a defense of the premise, cf. M. J. S. Hodge and D. Kohn (1985), in D. Kohn, ed. (1985, 205).

6. The standard example of Darwin's "shift" in assigning "explanatory weight" for various factors in his theory comes in *Descent of Man* (1871, 152–3): "In the earlier editions of the "Origin of Species" I probably attributed too much to the action of natural selection or the survival of the fittest [as compared with] inherited effects of habit [and] direct action of surrounding conditions." This statement (and similar ones, as, for example, in letters Darwin sent to St. G. J. Mivart in early 1871 [*CCD* 19, 30–9]) seems gratuitous. In larger context Darwin did not alter the priority of importance of explanatory factors: natural selection first, "chance" variations second, and inherited habit and "direct action" third. R. J. Richards (1989, 193–5 and n. 29) for discussion. David Hull (chapter 26 in D. Kohn, ed., 1985) provides a useful analysis of "Darwinism" as an ever-shifting historical construct that lacks an "essence" that many scholars have attempted to find, although even Hull appears to find two "essences" in Darwin's thinking about what "Darwinism" is: pluralism (i.e., the acknowledgment of multiple causes of variation) and Lyellian gradualism (800–9). No doubt many of the modern controversies about what "Darwinism" means center on these two factors. But Darwin also always insisted on the primacy of the mechanism of natural selection in evolutionary change and, as I argue here, on "chance" (properly understood) as the most important cause of variation. Recent defenders of the idea of a "Darwinian essence" include Gould (2002, Chapter 2); and Lennox, 2004.

7. R. J. Richards (1989), 90–1 cites F. Cuvier (1827) as a plausible initial source for Darwin's idea of "chance variation." But even if Darwin first discovered "chance variation" in Cuvier's "fortuitous modification, fugitive want, accidental habit" we must recall that Darwin would have to take these ideas further than Cuvier did (Cuvier was anti-transmutation) in order to get to his own theory of "indefinite modification" and "species change."

8. This is *Darwin's* core and would not necessarily be accepted by all "Darwinians," in his own day or, much less, in ours. But few self-described Darwinians of any age or stripe would likely find fault with this account of what Darwin himself believed to be the crux of his theory of evolution. A similar, if somewhat expanded reproduction of the core is Lennox (2004).

9. Some scholars find a decisive shift in Darwin's views about sexual v. asexual reproduction from his earliest musings about the species question, in which sexual reproduction was at first the basis of all variation, to a somewhat later view that asexually reproducing organisms could

also yield variations. Discussions include Hodge and Kohn in Kohn, ed., (1985); and Hodge, in Krueger et al., eds. (1987, vol. 2, chapter 10).

10. As early as 1838 (before he read Malthus) Darwin tried to set out the core of his theory in abbreviated form. It remains remarkably fixed and unvarying to this day. Darwin's first summary was in *Notebook* E-58 as follows: "Three principles will account for all: (1) Grandchildren like grandfathers [i.e., heredity]; (2) Tendency to small change <<especially with physical change>> [i.e., variation]; (3) Great fertility in proportion to support of parents [i.e., superfecundity]." Natural Selection is not mentioned. Darwin continued to attempt to capture the core both in the "Sketch of 1842" (*EBNS* 57), and the "Essay of 1844" (ibid. 133–4); and again four times in *Origin* (once at the end of Chapter IV on "Natural Selection," (*Origin* 270: IV. 384); once at the end of Chapter VII on "Instinct" (*Origin* 423: VII. 288); and twice in the final chapter called "Recapitulation and Conclusion," (*Origin* 719: XIV. 8; 758–9: XIV. 268–70). Darwin appears to have seen Natural Selection as an additional premise to the other three, from which he wished to derive another conclusion entirely—viz., the formation of new species through time (see Lennox, 2004). Perhaps Darwin's stylistically most brilliant exposition is the tidy summary at the end of Chapter VII of the *Origin*: "To my imagination [the beautiful changes and adaptations in nature] are all small consequences of one general law, leading to the advancement of all organic beings, namely multiply, vary, let the strongest live and the weakest die" (*Origin* 423: VII. 288). Here again the formation of new species is the consequent of the premises. Cf. S. J. Gould (2002); A.G.N. Flew (1956); Kohn (1980); R. Lewontin (1978; and R.J. Richards' note on Lewontin, 1989, 100 n. 97); and David Hull, (2001), 54–5 for other formulations of the "syllogistic core," all only variations on Darwin's formulations. Beatty (1984) provides an efficient summary of Darwin's idea of natural selection in the context of his "chance variation."

11. Hodge and Kohn, in Kohn, ed., (1985, 185–205), argue that "chance variation" did not make an appearance in the *Notebooks* until the relatively late *Notebook* N (1839). They connect its discovery by Darwin to Darwin's realization that selection under domestication, wherein chance variation provides the raw materials for the breeder's craft, must have an analogue in nature. I accept the latter part of this argument, but I find evidence in the *Notebooks* that Darwin contemplated the possibility of chance variation much earlier, as early as 1837. I also demur from their contention that Darwin did not have "his own theory" until 1839. His discovery of chance in 1837 or earlier was plausibly his basis for thinking he had a theory of his own prior to the encounter in late 1838 with Malthus. Cf. also Kohn, 1980; and n. 19 below. R. J. Richards (1989), 90–9 offers useful suggestions about "chance" in the development of Darwin's thinking, taking the question back to his encounter with F. Cuvier's 1828 "Essay on the Domestication of Mammiferous Animals," first cited by Darwin in the B *Notebook* (1836), but the focus here is on "changed environments" that come about "by chance" rather than changes in organic structures that are purely "by chance."

12. I examine the question in depth in chapter 8, "Darwin's Giraffes." Note 2 of that chapter documents recent authorities who have argued for a Darwinian shift to Lamarckism; note 6 gives a partial bibliography of recent studies of Lamarck in the context of Darwinian evolutionary theory.

13. So did F. Cuvier (1827), cited in R.J. Richards, (1989, 90–1). More on Darwin's possible sources in chapter 6 of the present work.

14. Whewell, in fact, despite his commitment to induction in science, could never give up a role for miracles in organic adaptations. Even natural laws could not do all the work necessary

to explain organic adaptations: "Nothing has been pointed out in the existing order of things which has any analogy or resemblance, of any valid kind, to that creative energy which must be exerted in the creation of new species" (1840, 2: 133–4). About chance Whewell had this to say: "The assumption of a Final Cause [i.e., designed purpose] in the structure of each part of animals and plants is as inevitable as the assumption of an Efficient Cause for every event. The maxim that in organized bodies nothing is *in vain*, is as necessarily true as the maxim that nothing happens *by chance*" (1840, vol. 1, Aphorism CV, xxxv, original emphasis). Herschel too needed "design," not chance. Darwin, as much as he may have followed Herschel and Whewell in other ways, could not go along with them here. Cf. M. Ruse (1975, 1978); Thagard (1977); Gildenhuys (2004); and chapter 3 of the present work for an examination of Darwin's debt to the philosophical ideas of John F. W. Herschel and William Whewell. Cf. also Hull (in Kohn, ed., 1985, Chapter 26); Schweber (in Kohn, ed., 1985, 47–9).

15. Creationists and intelligent designers are not often found defending natural selection as a mechanism for preserving and destroying whole lines of life, for the reason that, by definition, if the Creator (or Intelligence) designs it, it is perfect, just as the Creator intended. As regards *species,* this is no doubt their position. But at the same time Creationists and intelligent designers—at least some of them—do acknowledge *variation* among individuals in a species, and also acknowledge that unfavorable variations hurt survival prospects of some individuals in nature and that some favorable variations help to enhance survival prospects of individuals who have them (a review of some of the recent literature can be found in Freeman Dyson, *New York Review of Books*, October 4, 2001, 24–7; and R. Dawkins, 2006, 113–4). Darwin was well aware of this (cf. *CCD* vol. 6, 371 [April 12, 1857] in a letter to Hooker: "The most firm stickers for independent [i.e., intelligent] creation admit [variation]"). Thus, Creationists and intelligent designers *might* be willing to embrace natural selection as the mechanism of sortition for preserving and destroying individual variant types. This implies or entails no embrace of evolution of *species* in biological history.

16. Whatever Darwin may have believed in private or said in public, the "revolution" he wrought is generally believed to have removed "conscious or purposeful design" from natural processes. Alexander Rosenberg (1985, 246) sums up this way: "One of the most salient features of Darwin's revolution is that it ended forever the biological appeal of the argument from design, which founded the teleology of nature on the desires, intentions, and conscious designs of God." "Chance" in Darwin's thought (as "chance variation") has received scholarly attention in recent years, notably by Beatty (2006), who discusses chance in Darwin's study of orchid varieties; Lennox (2004), who associates "chance" with Darwin's attack on "special creationism"; and Millstein, 2000, 2011, who discusses "chance" in contemporary evolutionary theory more generally. Other references to this literature are given below.

17. As will be discussed later, most of the early scholarly reviews of Darwin's *Origin*, those appearing in 1860–1861, did not even notice the role of "chance" in the theory. I can find only one review in that first wave of critical commentary, that of F. W. Hutton, that notices the important role for "chance" (*The Geologist* 1860, 3: 464–72; reprinted in D. L. Hull 1973, 293–30; cf. Hull's "Comments," 300–1). Darwin's concern about the damage being done to the reception of his theory may have been prompted in part by this review, but more likely by letters he received from correspondents in the same period, especially those from Asa Gray and Charles Lyell.

18. Ruse, a proponent of the view that Darwin's religious views changed over the years in an "agnostic" direction, has strengthened his stance in more recent work, in his chapter to the

co-edited volume *The Cambridge Companion to the "Origin of Species"* (Ruse and Richards, 2009, 2): "In religion, [Darwin's trip in Wales with Adam Sedgwick in 1831] was important because Darwin's rather literalistic Christianity started to fade and he became something of a deist, believing in God as unmoved mover and that the greatest signs of His powers are the workings of unbroken law rather than signs of miraculous intervention." See also ibid., 8, for evidence that suggests Ruse believes Darwin remained a deist from then on.

19. Dawkins, 2006, Chapter 1, makes a good case that this often-quoted statement (and others by Einstein) does not show that Einstein was a "believer" in a Christian God, as much as many believe it does.

20. The "God" in question here is the nineteenth-century God of Victorian England—an omniscient, omnipotent, omnipresent deity who was widely believed to play a determining role in life's complexities, whether immediately or remotely. God was also generally considered to be "good," again in a mid-nineteenth century British sense. Darwin generally refrained from going into details about what the word "God" meant to him (he often confessed that he was simply "in a muddle" about so profound an issue), and to fill in all the details from historical context would take us too far afield. But cf. Ruse (various dates) and Dawkins (2006, Chapter 2) for ample discussions.

21. How should one interpret Darwin's unwillingness to state publicly what his views about "God" were? Even Emma, his wife, could not be sure where he stood on this key question (*CCD* 19, [1871] 106). One can only guess. One guess is that he had become an atheist from early on (1837?) and wished to hide this opinion from even his closest intimates. Another guess, perhaps more plausible, is that, after having given the question a great deal of thought, he decided any answer either way would be "unphilosophical," and so just set the question aside as insolvable (for example, *CCD* vol. 19, 551 in a letter to F. E. Abbott). I address the question more fully in subsequent chapters.

22. Auguste Comte viewed the development of knowledge as having progressed through three stages: theological, metaphysical, and positive. Darwin had read an extensive review of the first two volumes of Comte's *Cours de philosophie positive* (Comte, 1830–1842) in 1838 (see *CCD* vol. 2, letter to Charles Lyell, September 14, 1838; and vol. 9, August 1, 1861, n. 12).

23. At a recent conference on Darwin (San Diego, 2011), when I mentioned to the philosopher David Depew that I was working on "chance in Darwin's thought," he said "a complex topic. No, a *very* complex topic." I believed it then, and believe it even more so now.

24. An excellent review of the various meanings of chance in contemporary evolutionary theory is Millstein (2011, 425 ff). She identifies seven distinct meanings, all of which she attempts to subsume under what she calls a "unified chance concept." Three of these she traces back to Darwin himself, and of these three I shall be concerned mainly with her "sixth" meaning: evolutionary chance (variations leading to species changes that come about with no necessary connection to future adaptive needs).

25. Beatty (1984, 204–5) draws a contrast between "stochastic" (or random) and "deterministic" processes in evolution (cf. Richardson, 2006; Millstein, 2000, 2011; Eble, 2006; Grantham, 1999). His argument is that evolution is *always* a "stochastic" process (i.e., always involves randomness) because: (1) contrary to Hardy-Weinberg, populations in nature are never of an infinite size (in which case random shifts in gene-frequencies over time would "balance out" to no net changes); and (2) because "random drift" is always or nearly always *one* factor in the distribution of gene-frequencies in natural populations. He does not claim that "natural

selection" by itself (i.e., without any role for random drift) would be a "deterministic" process; he does not consider the case (but Richardson, 2006, 646, building on Beatty's work, makes the claim that it would be). But neither did Darwin. Beatty (and many others) is "Darwinian" in recognizing stochasticity in evolution. He departs from (or improves upon) Darwin's theory by: (1) attributing chance variations to "random drift" in gene-frequencies (Darwin was innocent of any knowledge of genes or "sampling errors" and so simply said "chance"); and (2) in arguing that "drift" and "selection" can and do act "concurrently" rather than "consecutively" in evolution. A more recent formulation is Beatty, in Krueger (1987, vol. 2, chapter 11); cf. also Lennox (2004).

26. A nice example from Darwin's day comes from William Thompson in his presidential address to the British Association for the Advancement of Science in 1871, in which he referred to Jonathan Swift's "absurd machines . . . that write books by randomly rearranging words" (quoted in *CCD* vol. 19, 526 n. 7).

27. James Lennox (2004) argues that Darwin's discussion of variation in his theory (in the sense of "random") was a response to *two* sets of criticism: those who wanted to see "design" in evolution, and those who followed Lamarck in seeing no room for chance in transmutation. "Apart from those urging Darwin to give up chance in favor of design, he had pressure to abandon chance from another direction, the evolutionary philosophy of Jean-Baptiste Lamarck. Lamarck's is another, materialistic argument against the variation in nature being a matter of chance. It is true that Darwin's theory could dispense with Lamarckian "adaptation of living beings to their changing environments" and also his "progressivism." But I do not find any "pressure" on Darwin from this quarter. The group he was most concerned with was the former—colleagues and friends (many of them) who wanted design. As to Lamarck, Darwin early on simply dismissed him and turned his back on his views, as most of the people he cared about the most had also done (see chapter 8 of the present work).

28. *CCD* vol. 13, 390 (supplement), [February 7, 1857] to B. P. Brent, where Darwin refers to his forthcoming species work as "my book on variation," underscoring the importance of this theme for his overall project. The editors of *CCD* (ibid., n. 8) claim that Darwin "sometimes" used the same expression in referring to this work (without citing where in the correspondence this could be found).

29. That chance might be thought to be the source of variation was a common worry in the 1860s and 70s: "To gaze on such a universe [as ours], to feel our hearts exult within us the fullness of existence, and to offer in explanation of such beneficent provision no other word but *Chance*, seems as unthankful and iniquitous as it seems absurd. . . . The hypothesis of Chance is inadmissible" (from *Contemporary Review* of September 1875, submitted by "P. C. W" and quoted in Asa Gray *Darwiniana*, 297–8; cf. John Beatty, 2006, 1–14 who analyzes the role of "chance" in Darwin's *Orchids*). For typical statements at the time about "chance" as the only alternative to design, cf. Paley's *Natural Theology*, 40–1, 281–7; Asa Gray's *Darwiniana*, 117, 125–26, 298 (Gray represents his own views and those of many other natural philosophers); *CCD*, vol. 8, 496 (from Darwin to A. Gray [November 1860]), vol. 10, 428 (from Gray to Darwin [August 1862]); vol. 11, 525 (from Gray to Darwin [July 1863]).

30. James Lennox (2004) makes the claim that Darwin *never* referred to variations as caused "by chance" in the *Origin*. Yet at the same time he allows that Darwin sometimes said selection operates on variations that "happen to occur [in nature]." I do not see the distinction, unless the only thing Lennox means is that Darwin did not use the *exact words* "variations occur by

chance." Certainly, variations that just "happen to arise" must be considered "chance variations" in Darwin's meaning.

31. Darwin sensed that some confusion as between "laws" and "chance" would result from his views when they were made public, as he confessed to his friend Hooker in 1856: "(No doubt the variability is governed by laws, some of which I am endeavouring very obscurely to trace).—The formation of a strong variety or species, I look at as almost wholly due to the selection of what may be incorrectly called *chance* variations or variability" (*CCD* vol. 5, 282 [23 November 1856], to Hooker, Darwin's emphasis). Sometimes Darwin used the word "accident" rather than "chance": "Hence, we may conclude that under domestication instincts have been acquired, and natural instincts have been lost, partly by habit, and partly by man selecting and accumulating, during successive generations, peculiar mental habits and actions, which at first appeared from what we must in our ignorance call an accident" (*Origin* 389: VII. 80). The idea is like that expressed at the beginning of chapter 5 of the present work: variations occur, by what process we know not, and so attribute them to "our ignorance." But this again refers to the *how* of variation, not the *what*, as I describe below. The only thing we can know, by observation, is whether the novelty is well adapted or not. If it is, it survives; if it is not, and the struggle with competitors is fierce (as it almost always is), it will perish.

DARWIN'S DICE

1

Two Faces of Chance

INTRODUCTION

In ordinary English usage the word "chance" is used in a number of different ways. Synonyms or words that may be substituted include accident, randomness, uncaused, fortune or fortuity, luck, happenstance, likelihood (or probability), unpredictability, coincidence, spontaneity, and serendipity, to mention only some of the most common. A similarly wide array of terms stands in direct contrast or opposition to "chance": determinism, necessity, caused, predictability, skill, free will, purpose, design, and uniformity, to name a few.[1] These terms, in turn, play a central role in what Gigerenzer et al. call "the most recalcitrant biological controversies of modern times: vitalism, mechanism, teleology, essentialism, and levels of organization and explanation" (1989, 123).[2] In a similar vein, John Beatty (1984, 183) has characterized the issue of the role of "chance" in evolution as one of the "liveliest disputes in evolutionary biology today."

Despite the array of terms and contexts, broadly, two applications stand out in Darwin's thought, each of which might be illustrated by examples drawn from non-evolutionary and other everyday contexts. With Darwin always in mind, I shall examine "chance as probability" in the first section, "chance as randomness" in the second section, and chance as normally understood among contemporary philosophers of biology in the third.[3] My chief concern will be with "chance as randomness," as described below, because that is the sense that mainly concerned Darwin.

An important caveat before we begin: while Darwin often used the word "chance" to explain the origin of variations, he also often indicated that what he really meant in such contexts is only "what we in our ignorance call 'chance'" (cf. n. 1 of this chapter). He thus implicitly suggested that much, if not all, that we attribute to chance in studies of nature might be eliminated if we knew more. If that were his position on the origin of variations it is hard to see why anyone who accepted natural selection would be worried about his theory. Claiming ignorance is only a deferral of providing an explanation. The explanation, when it came, might well be perfectly acceptable to those who rejected "chance." Few people do or should expect to see all of nature's hidden mysteries exposed in a single brilliant flash of light.

This is especially true of those who, like Charles Lyell and Asa Gray, wanted to see a role in Darwin's theory for a "creative force" or an "intelligent designer" in nature's productions. They knew how difficult it would be to provide definitive evidence for this view, but they could hardly fault Darwin for claiming simply "not to know." In Darwin's day it was perfectly acceptable to think about the scientific enterprise as the attempt to carve further and further into the domain of the unknown with the knife of scientific explanation. In the words of M. J. S. Hodge, "[in asserting that we do not know the causes of variation, and so say 'they arise by chance,' Darwin] was in conformity with the customary view that chance congenital variants . . . were not uncaused but, in accordance with the *commonplace* ignorance interpretation of chance, caused by unseen prenatal conditions" (n. 3 above, 242, emphasis supplied). This seems to be the accepted view even today. In the words of one recent scientist, "[chance] rationalizes the gaps left by our ignorance, an ignorance that may be assumed to be reducible without limit as research progresses" (Gigerenzer et al., 1989, 171).

But Darwin *was* criticized for invoking "chance" as a cause of variations (cf. Beatty, 1984, 183–5). The criticism, it seems, was well founded. Not only did Darwin not expect *ever* to be able to provide the needed explanation for variations, he also could not believe that the explanation, if it ever did come, could make any reference to a creative force or intelligent design. This was his decisive departure from scientists who otherwise were sympathetic to his theory. Even if variations are caused, and Darwin certainly believed they are, the causes are of such a kind that they ruled out any role for creative intelligence. The reasons for this will be more fully illuminated as we move along, but briefly, they were mainly based on his opinion that many if not most variations were either harmful or, at best, non-useful to progeny of parents that produced them. Darwin could not see "intelligence" in this brute fact of nature. To this extent his reliance on the idea of "chance" as the main factor in variation was just what frightened his sympathetic readers, and rightly so. They were not worried about his self-proclaimed ignorance but they were worried that the causes, once known, would rule out a role for God. And for Darwin, they did.[4]

CHANCE AS PROBABILITY OR LIKELIHOOD

Some organisms, according to Darwin, have a better chance of surviving than others (cf. Lennox, 2004). The reason for this is straightforward. To survive means to be equipped with the proper structures and behaviors (in general terms, "traits") to be successful within the environment into which one is born, or to which one migrates, or in one that changes. This is "adaptedness."[5] (The fact that the meaning is straightforward does not mean it is easy to determine in every case what the "adaptations" are or how, or even whether, certain traits contribute to survival prospects!)[6] It has nothing at all to do with what brought about one's adaptedness, only with survival prospects once structures and behaviors are already in place. The better one's adaptive fitness, the better one's chances—the greater one's probability—of surviving.[7]

"Probability" comes into play in any number of different situations in everyday life.[8] To modify an example from Aristotle's *Physics*, if one happens to run into someone at the marketplace to whom one owes money, even while not expecting to encounter the person there at that time, it is easy to say "I unexpectedly chanced to run into a person today at the marketplace to whom I owe money." (For unexpected events of this sort Aristotle employed what in English is often called "coincidence.")[9] This is just another way of saying the probability of running into this person was small, but obviously not impossible. Probability is a variable magnitude, ranging from "highly unlikely" to "very likely" (additional discussions in Dawkins, 2006, chapter 2; and Krueger et al., 1987, vol. 2, chapters 10–13). But at either extreme and at all points in between it is still "chance," in the sense that the outcome is not known in advance, and cannot be known with certainty. An outcome that is known in advance with absolute certainty is no longer by "chance" in this sense.

Other common examples are found in many games: what is the "chance" of blindly drawing three consecutive aces from a true deck of cards, or of rolling five "sixes" in a game of Yahtzee? The odds (chances) can be calculated with more or less precision, depending on the situation, but uncertainty in such cases cannot be eliminated. Apparently people are more likely to invoke "chance" as an explanation when the actual result deviates, from a statistical standpoint, from the predicted result, but that choice of words is only a matter of taste.[10] Any time a possible outcome is unknown in advance one may justifiably "explain" whatever outcome does occur as "due (at least in part) to chance."

"Chance" is often regarded by humans as "good" or "bad," less often as neutral, but this too is just a matter of personal taste. For whatever reason, when chance is rendered by the word "luck" it usually connotes "good" luck, unless it is explicitly prefaced by the word "bad" (or a synonym), whereas when it is rendered by "accident" it tends to mean "bad" luck. "I got lucky" almost always means the good kind of chance; the same for someone whose "luck runs out." "I had an accident" generally means something bad happened: a food spill, a traffic collision, a wrong turn. But these modalities are, can I say, accidental, matters of convention, and, as noted, the convention is defeasible by adding appropriate qualifiers ("bad luck," "happy accident," and so forth). Darwin, for example, often employed the word "accident" to describe outcomes that, from the standpoint of survival, were actually good. Strictly speaking, chance is good or bad only for the one who experiences it, directly or indirectly. When an organism has a "good chance of surviving," this does not mean "it is good for it to survive," except from the limited perspective of itself and of other organisms whose survival depend on its surviving. But even this formulation is misleading, except perhaps in the case of humans. Non-human organisms evidently do not consciously place a normative value on their survival or on anything else. A better formulation would be: if an organism is to survive in its environment, it will be "good" for it to have the features it chances to have—a utility test. Without them it will likely perish.

CHANCE AND SKILL

A chance event, whether one believes it to be good or bad in its result, does not rule out the possibility that the event is also influenced by "skill" or, more generally, "agency." Consider any standard elimination tournament, such as a tennis tournament (more details in Dennett, 1996, 52–5). Players advance by defeating opponents, one at a time, until the last one standing at the end of the final game is pronounced the winner. The tournament is set up in such a way as to minimize chance by maximizing the likelihood that the player who "really" is most skillful at tennis will win. But that result does not always happen, except, trivially, that winning all the matches is what it means to be proclaimed the most skillful. Unexpected contingencies—chance—can ruin the superior player's ascent to the top spot. On a given day, a less-than-best player can beat the best. To minimize the role of chance, tournament officials go to great lengths to ensure the tournament really is a test of skill—most obviously by making the rules and conditions equal for both players, but also by increasing the number of sets one needs to win to claim victory. A five-set victory is a better test of skill than a three-set victory. But even the most carefully enacted precautions cannot eliminate chance entirely. Suppose that the victor needs to win 501 out of 1000 sets to be the winner: chance still can—and does—play a role, in multiple ways. If one of the players, for example, happens to wake up with a head cold on the day of the deciding match, the domain of chance has suddenly expanded considerably at the expense of the domain of skill.

In most everyday situations, especially where competition, victory, and defeat are in play, skill and chance "compete" with each other for mastery. Most real-life situations are not organized competitions like tennis matches and card games, but they are still competitions. Biological organisms are constantly engaged in competitions of one sort or another—to survive usually means to outdo one's rivals in the struggle for existence. This, of course, applies to humans as well (humans are, after all, biological organisms). Competitions of skill and chance among humans almost define life in the realms of politics and economics, for example.

How much of a victor's success to attribute to skill and how much to chance is impossible to calculate with mathematical precision, but some have tried. Machiavelli, for example, famously calculated that "chance," what he called *fortuna*, when not opposed by human agency, accounts for approximately 50% of all competitive outcomes in the comedy of human politics. When "tamed" (a term that I believe was coined in this context by Ian Hacking, quoted in Gigerenzer 1989, xiii) by human skill (what Machiavelli called *virtu*), that percentage goes down in direct proportion to the skill involved (Machiavelli assigned no percentages here).[11] Machiavelli's calculations are really more like guesses—his "data" were more impressionistic than scientific—but his point is valid: *fortuna* rules much of human life, but its dominion may be countered in varying degrees by conscious attempts to defeat it, that is by cultivating skill. The same is true in the non-human organic world, except that non-humans often hone their "skills" unconsciously. The measure of their success is results, not intentions (cf. Dawkins, 1989, 4).

A modern example of how steps can be taken to "tame" chance when it is understood as probability is meteorology. Weather forecasters are supposed to give accurate predictions of tomorrow's weather, but they often fail to get it right, as everyone knows. So we blame the weatherperson. We should actually be blaming the element of what we call "chance" in weather events. Meteorologists (and many of the rest of us) hedge our bets by casting weather predictions in probabilistic language: the chance of rain on Tuesday is 70%. Modeling weather patterns ("skill" applied to weather forecasting) helps to improve the accuracy of such predictions but as of yet, the models have not succeeded in ruling out a role for chance. One can imagine that someday weather models (products of skill) will be so accurate as to diminish to a vanishing point mistakes in the predictions, but we are still far away from that prospect: too many variables are at play for all of them to be captured even in the most sophisticated models. A butterfly's flight in China, as is said, will influence the local weather in San Francisco. One might similarly imagine a day when a dice-throwing machine will be so fine-tuned as to make the prediction of the outcome of a simultaneous toss of five fair dice accurate every time, but we can guess this possibility is likely never to become an actuality. And even if a machine could master this skill, humans could not. Some events are too complex to eliminate from human skill the uncertain outcomes we designate in shorthand as due to chance.

It might fairly be said that *every* natural event contains an element of chance. We base our opinions about what will happen tomorrow on what happened before. Weather models are constructed from past weather events. Handicapping in the game Yahtzee—the points awarded for different configurations of dice tosses—is based ultimately on patterns of previous tosses, even if the long track-record of such tosses has permitted mathematicians to define a "science of probability" that seems to remove the need for history in making predictions. Had the history of dice-tosses (and similar events) not exhibited regular patterns over a large number of instances, statisticians would not have had any empirical basis for discovering the laws constituting the science.

Are all natural events then *also* products of some degree of skill? Except by definitional fiat it is improbable that this is so. At least some natural events may involve no skill at all. Machiavelli and Dante were concerned with human actors, and humans, while maybe not having a monopoly on skill, seem to possess the capability for acquiring it through deliberate effort to a much greater degree than other biological organisms. Socrates's definition of skill (*techne*) requires its possessor to be able to say what the skill in question is, how it is acquired, and how it may be taught. These definitional requirements almost necessarily restrict the domain of skill to the human species. But we may be reluctant to accept Socrates's criteria. Any learned or deliberately invented (as contrasted to instinctual) behavior might be counted as "skill." That definition might rule in higher primates, some species of birds, and other creatures, while still ruling out plants, bacteria, and infusoria. We may even say, metaphorically, that behaviors that permit or help permit organisms to survive in the competition for existence are "skills," even if such behaviors are purely instinctual. No bright lines surround skill. No matter. The important point for "chance as probability of survival in a context of competition" is that it be understood as a counterpoint to skill. The smaller the domain of skill, the larger the domain of chance. Skill and chance in this context seem to exhaust the possibilities for explaining who wins.[12]

CHANCE AND LAWS

Not all natural events, though, are competitions as we have employed the idea to this point, yet chance is still operating. Sometimes we just want to know what will happen in the future. Will the sun come up tomorrow? Will I get wet if I go out into the rain without an umbrella? Will the Illinois Central arrive at the station at precisely 2:10 p.m. as advertised? Our answers are all "guesses," but whether we guess correctly or not may have virtually no bearing on our survival prospects in a competitive environment. We may "guess" that the sun will not come up tomorrow, but our guess is not ventured with survival prospects in mind, even if what actually happens affects those prospects. (If the sun does come up we will survive for a while—probably—and if it does not we won't—much more probably.) Our confidence in our answers to these questions, as in answers to all questions about what will happen, depends upon history. But history is not authoritative when it comes to the future. Unanticipated contingencies may interfere with any natural event. The train may have broken down on its way to the station. The rain might mysteriously lose its wetness-causing property. Yesterday may have been the sun's last stand. We cannot know with absolute certainty, and so chance as "probability" remains an ever-present source of uncertainty in natural events.

Yet some probabilities are more probable than others. Gamblers are cautioned not to commit the "gambler's fallacy" (although many fail to heed the advice!). This is the fallacy that infers from past success at the roulette wheel to future success. You may think you are "on a roll," but in normal gambling situations that is never true, except in the sense that you have been lucky so far. Your luck can run out the very next time. But the advice is not invariably good. Sometimes the gambler's fallacy is no fallacy at all. If someone were to offer me an honest wager that the sun will not come up tomorrow, I would jump at the bet—and yet in doing so I would be in a technical sense succumbing to the gambler's fallacy. But that particular wager is no fool's game: I am virtually certain to win. This is not because I possess some secret knowledge about the sun's behavior or that I have devised a clever way to cheat. I will bet that the sun will come up because it is almost guaranteed on my current beliefs based on my past experience that it will. I *could* be wrong, true, but the *probability* is overwhelmingly great that I will be right. Some bets are better than others.[13]

What all of this goes to show is that chance is not "either/or." It is simply incorrect to say that sometimes things will happen by chance and other times they will not. All natural events are the result of a mix of both chance and non-chance factors (Beatty, 1984, as this applies to evolutionary theory). In consequence, guesses about the future may be placed along a continuum from "the extremely likely" to the "extremely unlikely." The further toward the latter end of this spectrum we find ourselves, the more likely (!) we are to invoke "chance" as the correct explanation. In truth, chance is in play at every station along the way.

If this is correct, and chance really does enter into the explanation of *all* natural events, in no matter how limited a way, then strictly speaking there are no natural laws according to the criterion of "invariable."[14] All "natural laws" are really just probability statements.[15] The reason is twofold. First, no two natural events occurring in space and time are identical in every respect. Every historical event is unique to some extent, no matter how small the difference from other like events it may be.[16] Because the discovery of natural laws is based on history, the idea of an invariable law is a contradiction. Second, it is humanly impossible exhaustively to describe *all* conditions influencing any event: remember the butterfly in China! If the hallmark of a "law" is to specify "invariable" necessary and sufficient conditions for an event, no laws will ever be established.

Nevertheless, as is true of the gambler's wager, some stabs at finding laws are better than others. Some events tend to recur more reliably and predictably than others. The domain of chance as "probability" is not infinite. By loosening slightly our standard for what a "genuine law" is, we can salvage the law-finding enterprise. A law subsumes a multiplicity of closely similar events under a single explanatory statement. If we allow all laws to be marginally provisional and contingent (i.e., the necessary and sufficient conditions to be acknowledged to be less than certain and exhaustive), then we may still proceed with the worthwhile attempt to discover natural laws.[17] A good test (not the only one) for identifying laws in this slightly restricted sense would be a probability test: what is the likelihood that a law succeeds in predicting correctly events of a certain kind that have not yet happened? Assuming we can cast the explanation for why the sun comes up every day when and where it is predicted to as a law, we would place the probability of its being a good law at a very high level. So far, predictions based on that law have not been wrong since the dawn of (earthly) time. But of course tomorrow is another day!

DARWIN AND PROBABILITY

When Darwin argued that some organisms have a better "chance" of surviving than others (a frequent expression in the *Origin* and elsewhere), he was tacitly acknowledging the points made in the preceding section: both that some events are more likely to occur than others, and that no event (in this case, survival of organisms) can be predicted with certainty (cf. Lennox, 2004).

Nevertheless, he did believe the patterns observed in nature are law-governed, and he also believed that these laws are "fixed," in the sense of unvarying and unexceptional. Thus, he departed from the view expressed above that no true laws, *stricto sensu*, exist. At the same time, however, he was close to the current view in thinking that actual observed phenomena do not always exhibit the patterns that would be predicted by the operation of unfailing laws. We should take a closer look at both sides of this: why Darwin believed that "fixed laws" do exist and perhaps sometimes can be found; and that natural phenomena often seem to defy the laws that are supposed to explain them.

As to the former, when Darwin first started thinking about a "theory" of evolution, he saw immediately that he would have to cast his theory in terms of "laws" if it were to have scientific respectability. But why should laws (i.e., statements of invariable relations among phenomena) exist at all? Why should not the phenomena of the universe be found to be irregular, uncertain, and even massively contingent—a grand accident? At the beginning of his studies Darwin rejected that view. The reason is that he believed, at least at first, that laws were the product of divine creation:

> The question if creative power acted at Galapagos it so acted that birds with plumage and tone of voice partly American North and South—**and geographical division are arbitrary and not permanent. This might be made very strong if we believe Creator creates by any laws, which I think is shown by the very facts of the Zoological character of these islands.** Such an influence Must exist in such spots (*Charles Darwin's Notebooks* (*CDN*) B-98, Darwin's emphasis).

Because laws were, he believed, divinely created, they must be perfect—not admitting of exception, even by God himself:[18]

> What a magnificent view one can take of the world—Astronomical causes modified by unknown ones cause changes in geography & changes of climate superadded to change of climate from physical causes...and these changing affect one another [by] certain laws of harmony keep perfect in these themselves

> [leading to] the Myriads of distinct forms from a period short of eternity to the present time, to the future. How far grander than idea from cramped imagination that God created [everything directly by immediate intervention], (warring against those very laws he established in all organic nature) (*CDN* D-36).

God's laws may not be easy to fathom, both as to what they are and as to why He created *these* laws and not others. The fact that mammalian destruction on a large scale is part of the natural order does not mean no laws are in play and also does not mean the law is not God's design:

> The extinction of the S. American quadrupeds is difficulty on any theory—without God is supposed to create and destroy without a rule—But what does he in this world without rule? The destruction of great Mammals over whole world shows there is rule (*CDN* D-72).

Attributing the origin of natural laws to a Divine maker would seem to rule out a role for chance. "There must be laws of variation, chance would never produce the feathers [of pigeons] or make [the variety of] pigeon breeds" (*St. Helena Model Notebook*, 48, quoted in *CDN* 362, n. 100-1; for reference, this *Notebook* is one of Darwin's earliest, 1836 or 1837 perhaps [Barrett, in Gruber, 1981]). When chance is excluded, law explains everything. Darwin devoted his scientific career to discovering these necessary laws of organic being. In time he abandoned the idea that God was responsible for legislating the laws of nature, or even, apparently, that God exists, but he did not stop believing that laws of nature exist. Late in life, in his *Autobiography*, he recalled that he had from early youth always been motivated by "the strongest desire to understand or explain whatever I observed, that is, to group all facts under some general laws" (1958 [1876], 141). Nor did he stop believing the laws of nature are exceptionless and uniform in operation. In the same section of the *Autobiography* where he confessed his fall away from Christian faith he wrote that "everything in nature [from the hinge of a bivalve to the course of a blowing wind] is the result of fixed laws [rather than intelligent or designed creation]" (1958 [1876], 87).

If that is so, why do we find Darwin increasingly insisting that natural phenomena are a result of chance? If exceptionless laws do exist, and one of those laws is natural selection (ibid., 87), why does Darwin say instead that the more fit have a "better chance" of surviving than the less fit? In saying this he seems to be acknowledging probability rather than certainty. Laws, especially fixed laws, should not admit of exceptions in the domain of their operation, whereas "better chance of survival" opens up a space for fortuity and accident.

The answer hinges on what Darwin took to be the value of discovering laws. Theories, in contrast to laws, are valuable just to the extent that they are able to "explain large classes of facts." He believed his theory of "descent with modification by natural selection" was a good instance of a theory. The test of a good law, by contrast, was different. A law is good to the extent that it enables accurate prediction about the future:

> Although no new fact be elicited by these speculation[s] even if partly true they are of the greatest service toward the end of science, namely prediction. Till facts are grouped and called there can be no prediction. The only advantage of discovering laws is to foretell what will happen & to see bearing of scattered facts (*CDN* D-67).

Predictions of natural events are never perfect, except possibly from God's point of view.[19] But that is not a reflection of the imperfection of laws, only of human understanding in grasping them. This is especially true of Darwin's "laws of variation," the subject of Chapter V of *Origin*. Again and again Darwin insists that these laws are obscure and often "unknown" (this will be treated in a later chapter). Much the same thing must be true of laws of "adaptive fitness." It is not the case that less-fit organisms might triumph over more fit ones, all things being equal. In nature "all things" covers too much territory to be comprehensively grasped by human intelligence. We should therefore regard the expression that "the more fit are more likely to succeed against the less fit in the struggle for existence" to mean something more like: "the more fit *will* triumph over the less fit; whenever a different result seems to occur it is only a reflection of our ignorance, not an exception to the law." "Survival of the fittest" in its actual operation, as opposed to what we might be able to discern about it, is as sturdy and fixed a law—that is, unexceptional—as one could wish for. If humans possessed divine knowledge they would discover in the law of natural selection no exceptions—or so, I will guess, Darwin believed.

CHANCE AS RANDOMNESS

Chance as "randomness" is a special instance of chance as "probability," but it requires separate treatment because it opens up a host of new issues, not treated above, that were of central concern to Darwin. The concept is operative in Darwinian theory primarily with regard to the causes of variation between adult organisms and their offspring. Whereas most naturalists of his day believed "like produces like," Darwin based his theory on the premise that like produces *unlike*.[20]

No offspring is identical to its parent(s). Darwin did not know exactly why this was the case, but he believed it to be indisputably so, and also indispensable to the theory of natural selection. Evolution could not occur if organisms invariably produced identical replicas of themselves. The variations need not be great; indeed, by virtue of *natura non facit saltum* (nature does not make jumps) they could not be great. But even the smallest variations—a grain of sand in the balance, as he once wrote—could and did add up to enormous changes in structures and behaviors in organic beings, given enough time. The question Darwin needed to confront to make his theory complete was, why do variations occur?

One common answer was: we do not know. Nevertheless, Darwin actively promoted causal explanations throughout his life. Early in his career he put the burden on "divine creation," but in time he came to regard that as a non-answer—a mere substitution of one inscrutable phenomenon by another even more inscrutable one. By the time of the *Origin* he eliminated that explanation (or better, actively refuted it) and replaced it by several others (examined in greater detail later; see chapter 3). The *Variation of Animals and Plants under Domestication* (1868) brought forward yet another, the doctrine of "pangenesis," always said by Darwin to be "only a hypothesis." But most commonly, from beginning to end, Darwin could often do no better than to say "variations are brought about by what we, in our ignorance, must call chance" (*Origin*, [*Variorum* 275] V, 4–5).

Whatever Darwin may have meant by this expression, one thing he did not mean was that variations are "not caused," or in the language of the preceding section, not outside of the dominion of law. "Every variation must have a cause" was a tenet of faith for Darwin (e.g., *Origin* [*Variorum* 320] V, 305). Some of the causes he suggested were genuinely "causal," in Aristotle's sense of an "efficient" cause. Factor "x" is just what brings about factor "y." Causes of this sort included (in *Origin*) the Lamarckian factor of use-inheritance, the "direct action" of conditions of existence on organization and behavior (sometimes called Geoffroyism, e.g., a colder climate causes longer fur to grow), and above all the "indirect action" of environmental conditions on the reproductive organs of adults prior to the conception of offspring. Other causes are better understood as "correlations" rather than causes. Examples of these include: correlations of growth (some changes in organization are always or usually accompanied by other changes, without necessarily causing them); species consisting of large populations display more variations than species consisting of small populations; and species whose members display much variation are more likely to undergo more variations through time than species whose members vary less.

Darwin was apparently in error about some of these causes or about some of their applications. Lamarckian use-inheritance as commonly understood is generally discredited in contemporary evolutionary science: giraffes did not acquire their long necks through deliberate stretching, to mention the iconic example (more on this in chapter 8 of this volume).[21] So is Geoffroyism, at least if it suggests that changes to structure/behavior induced directly by the environment are passed down through propagation (e.g., a suntan). Some of Darwin's standard examples of correlations of growth have also been found to be incorrect—his belief, for example, that cats with blue eyes are invariably deaf, and others.

But even when he was on the right path, Darwin was not always satisfied that he had given a complete account of the causes of variation. In these cases he always fell back, in apparent desperation, on the notion of "chance" as yet another cause. (An important issue will be to see how Darwin frequently reworked the way he articulated this "cause" over his long career.) Even when he asserted in writing that he only meant "cause unknown," he was putting his finger on a factor in variation that, even today, is best rendered by the word "chance." This is chance as "randomness."

Randomness in what sense? Certainly not in the sense of uncaused. Variations are caused, Darwin was sure, and he was no doubt right about that. Nor does it mean isotropy—the possibility of variation or organic change from one generation to the next in any direction and to any degree. The scope of possible variation, it is generally agreed, is limited by various "constraints," physical, biological, and historical (discussions in Gould, 2002; and Dennett, 1996).[22] But randomness has acquired a specific meaning in biological theory that has withstood all attempts to displace it: the idea that variations are not "pre-adaptations." This is the idea that creatures produce variations that are specifically designed with reference, conscious or unconscious, to the future adaptive needs of organisms that acquire them. Variations in this sense are truly "random." Some variations that happen to arise will be favorable for survival needs, others will be unfavorable, and still others, perhaps the great majority, will be "neutral," exhibiting characteristics that neither help nor hinder the survival of the organisms that possess them. In the latter case the neutral value of the variation will "cause" natural selection simply to "overlook" or "not see" the variation in question—and so, barring a later variation that cancels it out or later changes in the environment, it will persist in the population as a harmless guest.

This meaning of chance—chance as "variations that are blind (random) with respect to future adaptive needs"—is the core meaning that is the focus of the rest of this work. But even within this restricted scope we find several different variant usages. These usages come back ultimately to the core meaning, but each of them occupies a distinct place in contemporary Darwinism. Darwin himself is responsible for identifying some of them. Others have been discovered only with more recent advances of knowledge in evolutionary biology (above all genetics, about which Darwin knew nothing). But Darwin appears to be the first one to nail down correctly the overarching idea. "Randomness" in nature as the correct way to describe how variations come about, I venture to say, may have been Darwin's greatest discovery. And, as we shall see, it was also the idea that caused the greatest alarm among otherwise sympathetic readers. It connoted "undesigned," and that was the rub.

"CHANCE" IN CONTEMPORARY BIOLOGICAL THEORY

One may today enumerate a number of ways in which chance as randomness plays a role in evolution.[23] Some of these were anticipated by Darwin, but most were not. He did not know about Mendel, genes, or DNA, for example, whereas much of the current literature on chance in evolutionary biology is focused on these factors. Most scholars agree, however, that these discussions owe a fundamental debt to Darwin's pioneering thought about chance (e.g., Millstein, 2000, 2011; Beatty, 1984, 1987, 2006; Richardson, 2006; Eble, 1999).

Before I present my own encapsulation of current ideas, I should note that modern controversies about "chance" in evolution have tended to center on the idea of "random genetic drift" (discussions in Beatty, 1984, 1987; Millstein, 2000, 2011; and Turner, 1987). In this usage "randomness" is usually contrasted with "determinism," the idea that natural processes are constrained by biological necessity, even if the causal pathways are not yet fully understood. The "random" vs. "determined" distinction is not one we find in Darwin's thought as such. He certainly did not know about "drift" or "random walks," for reasons already stated. But he did foreshadow it, in the sense that "natural selection" was to him "deterministic," whereas "variations" are not (see previous two sections of this chapter). I therefore surmise that, equipped with current knowledge and confronted with the following list of "chance" factors in evolution, he would recognize his own hand in shaping it.

(a) We cannot at present know with certainty when, why, or where on a chromosome a mutation may occur. Mutations seem to occur "by chance," or randomly.

(b) We cannot know with certainty when, why, or where a gene crossover or genetic recombination may occur, and so we say these events occur "by chance," or randomly.

(c) We often cannot know, except retrospectively, what phenotypic effects, if any, changes in genes or gene configurations will have on organisms. These effects may then be said to come about "by chance."

(d) Changes to organisms are "blind" with respect to future adaptive needs. Nature does not "look ahead" when it produces variations. Some variations just happen to be favorable, others unfavorable, and others neutral in value for organism survival. Variation may then be said to be "random," or "by chance."[24]

(e) "Gene frequency changes" from one generation of organisms to the next brought about by random "sampling errors" may continue to accumulate over the generations, even where no selective value to the organisms possessing them is operative. "Gene frequency changes that occur in this way are referred to as 'random drifts' in gene frequency" (Gigerenzer et al., 154; cf. Beatty, 1984; 2010; Hodge, 1987, chapter 10; Beatty, 1987, chapter 11).

(f) Gene frequencies of "neutral" value for survival needs of organisms may be *perpetuated* by "chance," in the sense that they arise by random drift and are thereafter overlooked by the directing agency of natural selection.[25]

(g) In general, for all the above reasons, variations are unpredictable, at least in current understanding. What comes along just comes along, and so we say the variations come along "by chance."

"Chance" enters evolutionary biology in yet other ways:

(h) Life itself originated by an accidental coming together of the right materials in the right environment. Life is thus ultimately "by chance."[26]

(i) Geographical distribution of plants and animals around the globe is often due to what Darwin called "chance transport." A seed clinging to an animal's fur being thus transported to a new location is one example. A bird being wafted by strong wind currents to a new location is another. Darwin names many other examples. "Chance transport" as a factor in biogeographical distribution is the subject of chapter 2 of this work.

Even the beginnings of human civilization, which Darwin associated with the development of agriculture, can be attributed to chance:

(j) "[The habits of civilized man] almost necessitate the cultivation of the ground; and the first steps in cultivation would probably result, as I have elsewhere shewn, from some such accident as the seeds of the fruit-tree falling on a heap of refuse and producing an unusually fine variety" (*Descent*, Part I, Ch. 5, 167).[27]

RECENT EXPOSITIONS

A classic exposition of randomness at the level of DNA—(a) and (b) above—is Jacques Monod's work, first published in French in 1970 (translated into English in 1971) in the book *Chance and Necessity.* The issue, as Monod represents it, concerns the necessary fact of copying errors in the process of DNA replication. Like most copying errors, no rhyme or reason can be assigned for why certain errors are made and not others. They are "by chance," and not just because we are ignorant. No amount of knowledge, Monod suggests, can ever enable us to understand the causes or predict the effects of replication errors when they occur, other than simply knowing that they have occurred after the fact. If science aims at prediction (among other goals)—indeed, if the possibility of making predictions is one of the hallmarks of science—then evolutionary theory will never achieve the status of a complete science because transcription errors are *in principle* unpredictable.[28] Monod puts the point this way:

> We call these events [viz., DNA replication errors] accidental; we say that they are random occurrences. And since they constitute the *only* possible source of modifications in the genetic text, itself the *sole* repository of the organism's hereditary structure, it necessarily follows that chance *alone* is at the source of every innovation, of all creation in the biosphere (Monod, 1971, 112, emphasis in the original).[29]

The *reason* Monod advances for insisting on this point involves an analysis of the process of DNA replication and of how copying errors generate "functional effects." These effects, he argues, are so far removed from the actual mutational event and complicated by so many intermediating processes, that even were we able to predict mutational events before they happen—and progress along these lines has been made—we would not be able to know until after the fact what functional effects these mutational events, if any, will have. So remote are effects from causes that Monod speaks of "complete independence" between the two (ibid., 114). Where effects cannot be linked even in principle to causes, one must give up the hope of eliminating the role of chance from evolution. "In effect natural selection operates *upon* the products of chance and can feed nowhere else" (ibid., 118).

More recently, the philosopher Daniel Dennett, quoting the biologist Mark Ridley, expressed much the same idea, adding an appropriate caveat at the end:

> Ever since Darwin, [evolutionary] orthodoxy has presupposed that all mutation is random: *blind* chance makes the candidates. Mark Ridley provides the

> standard declaration: "Various theories of evolution by 'directed variation' have been proposed, but we must rule them out. There is no evidence for directed variation in mutation, in recombination, or in the process of Mendelian inheritance.... [Such] theories are in fact wrong." But that is a mite too strong. The orthodox theory mustn't *presuppose* any process of directed mutation... but it can leave open the possibility [of] nonmiraculous mechanisms that can bias the distribution of mutations in sped-up directions (Dennett, [1995], 323).

The basic idea is that phenotypic variation in organisms occurs as a result of changes in genetic configurations. How the latter come about and the changes they produce are not and perhaps cannot be fully understood. A convenient shorthand is that these variations are due to "chance."

Chance as "blind with respect to future adaptive needs"—(c) and (d) above—is well spelled out by the paleontologist S. J. Gould:

> If only some offspring can survive... then, on average, (as a statistical phenomenon, not a guarantee for any particular organism), survivors will be those individuals that, by their fortuity of varying in directions most suited for adaptation to changing local environments, will leave more surviving offspring than other members of the population (Gould, 2002, 126, n. *).[30]

It is important to emphasize here that "fortuity" implies "chance" only with respect to conditions in which organisms happen to find themselves. Since local conditions change, today's "good fit" may be tomorrow's "bad fit." And since organisms vary from generation to generation, a change in fortune may also result. In both cases Nature simply creates changing conditions and variations, and then, let come what may. As Gould explains:

> Darwinians have never argued for "random" mutation in the restricted and technical sense of "equally likely in all directions," as in tossing a die. But our sloppy use of "random" ... does capture, at least in a vernacular sense, the essence of the important claim that we do wish to convey—namely, that variation must be *unrelated to the direction of evolutionary change*; or, more strongly, that nothing about the process of creating raw material biases the pathway of subsequent change in adaptive directions (Gould, 2002, 144).

"Genes have no foresight," as Richard Dawkins puts it (Dawkins, 1989, 8, 24). This fact is one source of the image employed by Dawkins (and others) of "blindness" in the course of evolution.[31] Dennett expresses the idea this way:

> The Local Rule is fundamental to Darwinism; it is equivalent to the requirement that there cannot be any intelligent (or "far-seeing") foresight in the design process, but only ultimately stupid opportunistic exploitation of whatever lucky lifting happens your way (Dennett, 1 995, 191).

Blindness and stupidity are not always paired in the history of ideas (consider the blind priest Teiresius of ancient legend and theater), but in evolution they are usually found to walk hand-in-hand. This does not mean, however, that all variations are equally likely to occur, "equally likely in all directions," as Gould expressed it above. In Dennett's words:

> Among the many non-actual possibles, some are—or were—"more possible" than others: that is, their appearance was more *probable* than the appearance of others, simply because they were *neighbors* of actual genomes, only a few choices away in the random zipping-up process that puts together the new DNA volume from the parent drafts, or only one or a few random typos away in the great copying process (Dennett, 1995, 125, emphasis in the original; cf. Dawkins, 1989, 31).

David Hull (2001, 54–6) has urged evolutionary biologists and philosophers of biology to eschew any qualifying adjective for "variation" in biological evolution. "Blind," "random," "chance," or what have you, are "extremely misleading" terms, insofar as they cannot be accepted as literally true. Variations are *caused*, he notes. No one working in this area doubts this, and neither did Darwin (cf. *Origin* [*Variorum*] 320: I. 305; Rosenberg, 1985, 216–7). "Chance" and associated ideas might seem to suggest otherwise.

But to say all variations are caused is not to say that they are all known—the precise causal sequence giving rise to them—or even that all of them might someday be known. Some are known, and some are not. As Rosenberg writes, "Mendelian randomness…does not suggest that the segregation of genes is causally undetermined. Rather it is predicated on the notion that the causal determinants of [variation], like those involved in the rolling of fair dice, are inaccessible to practical genetic inquiry" (Rosenberg, 1985, 216–7; Hull, 2001, 56). If Monod is right, the full story of the precise interplay between DNA copying errors and consequent variations will never be known. Hull's advice is good advice, for the reason he gives, but it does not undermine the usefulness of "chance" as shorthand for "unknown and undesigned causes." That, in fact, is just how Darwin viewed the matter.

Causes always coexist with constraints, as they are often called, and constraints of various kinds may "bias the pathway" of evolutionary development. But what they cannot do is to ensure that only favorable variations will arise. No matter how many possible pathways are blocked off by laws of physics and chemistry, by biological complexities, and by historical contingencies, innumerable other pathways remain open. Most of these possibles will never become actual, and the ones that do will come about as a result of random sorting and sifting that, in the present and foreseeable state of human knowledge, can only be attributed to chance.[32]

It is also important to emphasize that the possible variations that do happen to arise are not necessarily beneficial to those organisms possessing them. They may be of neutral value, or harmful. To adapt an Aristotelian expression from ethics to a biological context, there are far more ways to be wrong than right. As Michael Ruse has recently written:

> What about more systematic failures of adaptation and design? At some basic level, this comes with the territory, for it has always been emphasized, from Darwin on, that the building blocks of evolution—the raw variations—do not come (as Asa Gray supposed) according to need but purely by chance, in that they are not designed for immediate use. Hence, in some sense adaptation is always a struggle, no matter how efficient the power of selection (Ruse, 2003, 199).

Even the origin of life itself—(g) above—seems to be the product of an accident of near-miraculous proportions. Different scientific accounts of how this first beginning may have come about have been proposed, but virtually all of them, and certainly the most plausible among them, assign a significant role to random accidents that occurred in the primal soup, and that these were necessary for life to begin. Some of them enabled replication to occur (a requirement for evolution), while others by chance did not (cf. Kueppers, 1987, vol. 2, chapter 13). Dawkins, without going into details, sums up this way:

> Let me end with a brief manifesto, a summary of the entire selfish gene/extended phenotype view of life. It is a view, I maintain, that applies to living things everywhere in the universe. The fundamental unit, the prime mover of all life, is the replicator. A replicator is anything in the universe of which copies are made. Replicators come into existence, in the first place, by chance, by the random jostling of smaller particles. Once a replicator has come into existence it is capable of generating an infinitely large set of copies of itself. No copying process is perfect, however, and the population of replicators comes to include varieties that differ from one another (Dawkins, 1989, 264; a more extended account is in Dawkins, 1986, 148–58; cf. Monod, 1971, 97–8; and Dennett, 1996, 156–9).

Dawkins has applied this idea of "random jostling" to the entire process of evolutionary history in his more recent writings. In "Universal Darwinism" (Lewens et al., eds., 1998, 23–4), Dawkins drew an important distinction between big "saltational" variations and small incremental changes in design. The former are ruled out as the source of evolutionary change: even "chance" cannot produce successful "monsters." Natural selection only works on small changes. But these small changes—variations—must, given current lack of knowledge, be attributed to "chance:"

> The great virtue of the idea of evolution is that it explains, in terms of blind physical forces, the existence of undisputed adaptations whose statistical improbability is enormous, without recourse to the supernatural or the mystical. Since we *define* an undisputed adaptation as an adaptation that is too complex to have come about by chance, how is it possible to invoke only blind physical forces in explanation? The answer—Darwin's answer—is astonishingly simple when we consider how self-evident Paley's divine Watchmaker must have seemed to his contemporaries. The key is that the co-adapted parts do not have to be assembled *all at once*. They can be put together in small stages. But they really do have to be *small* stages. Otherwise we are back again with the problem we started with: the creation by chance of complexity that is too great to have been created by chance (Dawkins, in Lewens et al., eds. 1998, 23–4, emphasis in the original).

This passage, out of context, may seem to imply that Dawkins does not really believe variations are by chance at all. But clearly what he is ruling out are only *large-scale saltations* by chance.[33] The small changes upon which natural selection works are certainly by chance, or as Dawkins says, by "luck:"

> To repeat, the key to the Darwinian explanation of adaptive complexity is the replacement of instantaneous, coincidental, multidimensional luck by gradual, inch by inch, smeared out luck.... Darwinism—the non-random selection of randomly varying replicating entities by reason of their "phenotypic" effects—is the only force I know that can, in principle, guide evolution in the direction of adaptive complexity (Dawkins in Lewens, Hull and Ruse, eds. 1998, 24, 32).

While Dawkins emphasized the "inch by inch" character of chance in evolutionary history, Dennett wanted to show its ubiquity in time and space, an idea he captured in the memorable phrase "universal acid." Like a strange chemical concoction that is capable of eating through "everything it touches," Darwinian theory (i.e., chance variation and natural selection) is capable of "eating through," or explaining *all* natural organic productions, from the first origin of life (two amino acids happening to drift together three and a half billion years ago that fortuitously enabled replication to occur or some other fortuity) to the evolution of eukaryotes, then to more complex organic structures, all the way down to all of earth's organic diversity. And not only on earth. Dennett surmises that if organic life exists anywhere else in the universe it must have been produced by a similar sequence of random variation events followed by natural selection. Darwinism, with its foundational reliance on "chance variation," aided and abetted by natural selection, works to explain everything, as Dennett would say, "all the way down and all the way up."

With this background of understanding in general terms, we may now look more specifically at applications in Darwin's writing. From this point on I shall be focusing on "chance" as "random variation," as described above. Let us begin with a look at Darwin's first steps down this road, the early *Notebooks* (1837) and his ideas about "chance transport" in biogeographical distribution.

NOTES

1. The categories are not always stable, even in Darwin's writing. For example, in the *Notebooks* he claimed that "chance" and "free will" are synonymous (*CDN* M28–30; M72–3); and when he spoke of "chance variations," he insisted that this expression should not be taken to mean "not caused," but rather only "cause unknown" (*Origin* [*Variorum*], 275: V. 4–5). "Accident" and "spontaneity" were the terms most commonly employed by Darwin as synonyms for "chance."

2. Gigerenzer et al., 1989 is a useful introduction to the history of "chance" in a variety of disciplinary domains in science. Beatty (1984), Lennox (2004), and Millstein (2011) provide helpful surveys of Darwinian "chance variation" within a larger context of "meanings of chance in evolutionary biology."

3. Cf. M. J. S. Hodge, in Krueger et al., vol. 2 (1987), 233–70, who develops the distinction between "Darwin's 'chance'" (i.e., chance variation), and "Darwin's 'chances'" (i.e., chances of survival of organisms in nature). This distinction nicely illustrates the "two faces of chance" in the title of this chapter.

4. Darwin's theory ruled out a role not only for an intervenient God in nature's productions, but more generally, a role for traditional teleology (everything in nature is made for a purpose) and for natural theology in its Paleyan form (intelligent design or special creationism). Cf. Beatty (2006) and Lennox (2004) for discussions.

5. A distinction between "adaptedness" and "fitness" is drawn by Hodge (n. 3 above, 257–9). Adaptedness is the result of a "causal process" whereas "fitness" is better understood as an "expectancy," and expectancies are themselves not causal and so do not provide explanations. More on this in chapter 3 of this volume.

6. A robust literature about this subject has sprung up in recent years, provoked in part by the controversial proposal of S. J. Gould and R. Lewontin that the search for adaptive fitness for all structures and behaviors is misguided. Many structures and behaviors, they claim, have no adaptive value and so to search for them is to engage in the game of inventing "just-so stories" about

why and how they work to ensure survival. The dispute resembles one that took place much earlier in the twentieth century between proponents of "random drift" and "selectionists." The former believed that changes in gene frequencies in populations could come about through drift, even where no selective advantage was bestowed, whereas the latter did not. Discussions include Gigerenzer et al. (1989, 154–9); Beatty (1984); Dennett (1996, 267–81); Hodge (in Krueger et al., vol. 2, 1987, 250–66); Beatty (in Krueger et al., vol. 2, 1987, 271–311); and Turner (in Krueger et al., vol. 2, 1987, 313–54).

7. To say that organisms that survive are well suited to their conditions "by definition" does not mean, as some have argued, that natural selection is a tautology. The issue has been amply discussed in Sober (2000), Hull and Ruse (1998), Mayr (1982), Hodge (in Krueger et al., vol. 2, 1987), 257–66; and Brandon (1996).

8. Cf. Krueger et al. (1987), for a collection of essays on the development of the idea of probability in the natural sciences; vol. 2, section V, treats specifically the science of evolutionary biology.

9. Aristotle gives an extended analysis of "chance" as a factor in natural events in *Physics* Book II. He asks whether "chance" should be considered a "fifth cause" in natural events (in other words, whether any natural events are "uncaused" in the usual senses); and, relatedly, whether the structures of biological organisms have come into being "by chance" rather than "by purpose or necessity." He answers both questions in the negative. (The philosopher Empedocles, Aristotle believed, gave an affirmative to the second question. In the "Historical Sketch" at the beginning of the *Origin*, Darwin seems to believe the Empedoclean view was actually Aristotle's [Johnson, 2007]). Aristotle's employment of "coincidence" to render events such as "chance encounters" permits him to show, he thinks, that such encounters are really not "by chance" but may be explained by other "causes."

10. Gigerenzer et al. (1990) suggest that human calculations of probability in betting situations (a category that covers a surprisingly large range of instances) is generally to be construed as a "risk-reward" calculation: the greater the risk, the greater the reward must be for rational agents to enter into the wager. The *locus classicus* is "Pascal's wager": it is rational to believe in God because the rewards for believing, though remote, are infinitely good, whereas the risks for not believing, though uncertain, are infinitely bad (1 ff.). Dawkins (2006, 103–5) provides some criticisms and counter-arguments.

11. A parallel exercise has recently been undertaken by the Bayesian economist Stephen Unwin in *The Probability of God* (2003), who has determined that the probability of "God's" existence is 67%, even before factoring in "faith." When faith is factored in, the probability zooms up to 95%. Dawkin's (2006, 105–8) gives a delightful discussion and critique of this method of "proof" when applied to God. The issue is relevant here because if God does exist (and "rules the universe, is omnipotent," etc.), any role for "chance" in nature goes down to zero, except in the sense of "what we in our ignorance call chance."

12. I am deliberately sidestepping the "conundrum" of "determinism" and "free will" here, which begs the question: is the best player "predestined" to win (or lose)? Maybe God wanted him to win; and did the best player win (or lose) by virtue of a notoriously unpredictable "free will"? Maybe he chose to stay home on the day of the match or deliberately double-faulted on match point. On either account neither skill nor chance is the right explanation for victory (or loss). The question when applied to studies of nature was important to Darwin, but it cuts in a different way than the "chance/skill" distinction under consideration here. The question is of sufficient importance to merit separate treatment in chapters 9 and 10.

13. A corollary: betting *against* the sun coming up tomorrow would also not count as a commission of the gambler's fallacy. The sun has always come up the next day, so the prediction that it will not come up tomorrow is based on no track record of history. Such a bet invites a new category of explanation: stupidity. But then again, some "day" it will not come up.

14. An ambiguity in Darwin's usage of the word "laws" (often rendered by him as "fixed laws") must be briefly noted here (explored in greater detail in a later chapter): sometimes a "law" is a "statement" that affirms a regular, even unchanging sequence of events in nature. For example, the "law of gravity" describes in words the unvarying attraction of bodies to one another in terms of distance, mass, and so forth. At other times a "law" is the actual force that makes natural events behave the way they do. On this sense, the law of gravity is not a statement of physical relations or sequences of events but a "power" that "governs" natural events. Darwin apparently did not notice the ambiguity, but as a rule used the term in the former rather than the latter sense. Unless otherwise noted I shall be assuming the former sense as Darwin's intended meaning.

15. Philosophers sometimes refer to laws as statements of all conditions necessary and sufficient for invariably bringing like phenomena into existence over an indefinitely repeated number of like instances. This could change. Not too long ago the word that was used in place of "necessary and sufficient conditions" was "cause." David Hume showed why this was a bad way of explaining natural events: causes cannot be "seen" or "observed." They are inferences based upon the observation of phenomena that can be seen. "Necessary and sufficient conditions" does not commit the scientist to having to posit unobservable entities in explanations. "Conditions" of both sorts are observable. Hume's formulation is no doubt an improvement, but one cannot rule out the possibility that a new, improved way of formulating laws will be devised at a future time. In keeping with the present theme, we may say that the "likelihood" of such a development in scientific epistemology may be small, but who would dare to say impossible? Cf. M. J. S Hodge, in Krueger et al. (vol. 2, 1987, chapter 10) for a discussion of issues concerning "laws" and "causes" in evolutionary biology.

16. This is the reason why Karl Popper's criterion of "falsifiability" for identifying genuine scientific statements has been criticized: it is impossible to say that the apparent falseness of any empirical statement that goes into the construction of a "law" is the *reason* the law is false. Too many other factors are involved (most of which are unstated or even unnoticed) to say that one causal factor is responsible for the falseness of the law. See Hodge (above, n. 3) for a more complete discussion.

17. Philosophers of biology sometimes question whether any "genuine" laws of biology exist. The reason, in many cases, comes down to their acceptance of a rigid understanding of what a "genuine" law is, or what conditions must be satisfied in order for an empirical proposition to count as a law or a constitutive part of a law. This understanding derives from a paradigmatic definition of laws borrowed from so-called "harder" science, especially physics. No doubt a gap exists in epistemic probability between "laws of physics" and "laws of biology"; the latter are dubious as "laws" because of the large number of "fortuitous" and "contingent" factors in their construction. But this gap is a matter of scientific practice, not principle. We may attempt to bridge it by suggesting that no laws of physics are perfectly "certain" or "true," and that some biological "laws" are more probable than others. Statements about nature are neither "true or untrue" but only "more or less probably true or untrue." Further discussion in Dennett (1996).

18. Dawkins (2006, 58) observes that in the view of some "theists," in Darwin's day as in our own, the idea that God could not act against His own laws was an incorrect view. An all-powerful God can, by definition, do anything. This is a decidedly Hobbesian view of sovereign power, but even more Hobbesian than Hobbes, because Hobbes did put restraints on the Leviathan, or at least gave citizens "rights" against it (but not many). The central point, though, is correct: sovereigns, by virtue of being sovereign, are "above the law" for the simple reason that they make the law.

19. Even in 1837, when he believed that God was the divine legislator of natural laws, Darwin was not sure God could predict how those laws would play out in biological history: "How does it come wandering birds such [as] sandpipers not new at Galapagos. Did the creative force know that these species could arrive—did it only create those kinds not so likely to wander[?] Did it create two species closely allied to Mus. Coronata, but not coronata?" (*CDN* B-100). Even to ask such questions was implicitly to question the Christian orthodoxy of divine omniscience in creation. Many years later, Darwin showed even further departure from the faith: to Asa Gray in 1860 he declared that he had come to the conclusion that, even if the laws of nature had been divinely fashioned (about this he was agnostic), once God had made the laws he let them "play out in actual events according to what we must call chance" (*Correspondence of Charles Darwin*, hereafter *CCD*, 224 [May 24, 1860]).

20. I am not suggesting that Darwin was the only naturalist of his day to recognize variation; that was a commonplace. The difference between Darwin and the others was that while they believed variation had limits—an invisible cell wall beyond which they could not vary further—Darwin saw each variation as the potential raw material for a permanent modification that, with the help of natural selection, could become, in turn, the starting point for yet more variation, and so on indefinitely.

21. Lamarckism is apparently making a comeback in evolutionary thinking. For recent discussions see the volume edited by Pigliucci and Mueller (2010).

22. Michael Ghiselin makes a similar point in emphasizing the interconnectedness of parts of organisms and the consequent constraint this places on the pace and direction of evolutionary change: "Such constraints definitely tell us that the genetical changes that permit evolution are not 'random' in the sense of a phenotype with one genetic basis being just as probable as any other" (1997, 157).

23. The following list of usages/meanings is not intended to be exhaustive or systematic but rather to illustrate how extensive the idea has become in the days since Darwin. In particular I make no mention here of the "chances of survival" idea discussed in the second part or of the "chance survival of neutral variations" idea that has emerged in the neutralist literature. Cf. M. J. S. Hodge (in Krueger et al. 1987, vol. 2, chapter 10) and Lennox (2004) for reviews. Beatty (1984, 186–7) discriminates among "chance" as "undesigned," "chance" as the Aristotelian idea of "coincidence," and "chance" as the Laplacean idea of "cause unknown," and contrasts these with Darwinian "chance (i.e., random) variation" and the more recent idea of "chance" as "random drift" (cf. Shanahan, 1991). Eble (1999, and discussion in Millstein, 2000) identifies five meanings of chance in evolutionary biology, discriminating, importantly, between "chance as ignorance of causes" (a Laplacean idea) and "chance as random with regard to future adaptive needs" (the Darwinian idea under scrutiny here). Millstein (2011, 245–6) gives a "list" (with explanations) of seven "meanings of chance" in evolutionary theory as of the early twenty-first century. She traces only three of these back to Darwin. But these scholars

agree that Darwin is responsible for opening up the possibility that "chance" could be a factor in evolution at all.

24. D. Futuyma (1979, 240) gives the standard textbook description that combines (a) and (d) above: "Mutation is random in the *chance that a specific mutation will occur is not affected by how useful that mutation would be*" (quoted in Gigerenzer et al., 1989, 152, emphasis in the original; cf. also 138). This meaning corresponds to Millstein's (2011) sixth "meaning of chance," what she calls "evolutionary chance."

25. "... the great majority of evolutionary changes at the molecular (DNA) level do not result from Darwinian natural selection acting on advantageous mutants but, rather, from random fixation of selectively neutral or very nearly neutral mutants through random genetic drift, which is caused by random sampling of gametes in finite populations" (Kimura, 1992, 225). Lennox (2004) comments: "Here, it will be noticed, the focus is not on the *generation* of variations but on the *perpetuation* of variations. The contrast is between a random sampling of gametes that leads to the fixation of selectively neutral alleles and natural selection favoring advantageous variations. That is, the contrast between 'chance' and 'fitness biased' processes is now being used to distinguish *means of perpetuating certain variations*. We are contrasting two sampling processes. Drift samples without concern for adaptation; selection samples discriminately on the basis of differences in fitness. Both samplings are 'probabilistic', of course, but that in no way obviates the above contrast."

26. Darwin claimed in no uncertain terms that he was not concerned with the question of the origin of life: "I have nothing to do with the origin of... life itself" (*Origin* VII, 5 [*Variorum* 380]); cf. *CCD* vol. 19 (1871), 53, letter to J. D. Hooker. Dennett (1995, 156–63), reviewing several recent accounts, discusses the philosophical issues regarding chance in the creation of life and shows how it is possible to construct a plausible view about how life could have originated "spontaneously" (a Darwinian expression, as we shall see), that is, without divine intelligence. A basic theme of his book is that intelligence is a consequence, not a cause of evolution by natural selection. Cf. Dawkins (1989, 14–18; 2006, 137) and Kueppers (1987, chapter 13).

27. Other formulations may be found in Mayr (1963, 1988), Gould (2002), Dawkins (1989), Hull (2001), and Ghiselin (1997).

28. Kueppers (1987, 357–63) has challenged Monod on the "*unpredictable in principle*" notion of how variations arise, but he seems to agree that to get to the requisite knowledge for making accurate predictions about replicative variation would require more knowledge than we currently possess or even foresee.

29. Ernst Mayr (1963) had already postulated the "unpredictability in principle" of new variations (reasserted in Mayr, 1988, 475), but he did not give a principled argument, so to speak, for this assertion. That is something Monod was perhaps the first to supply.

30. Michael Ghiselin (1969, 165) also used the word "fortuitous" to give a Darwinian account of the cause of variation, but in contrasting it explicitly with "randomness" and associating it with embryology, he seems to be using the word in a more restricted sense. "A major conclusion derived from Darwin's embryological analysis of variation was the discovery... that variation is not random but fortuitous: it occurs according to its own laws, and these are basically those of developmental mechanics." Cf. also Ghiselin (1997, 156–7).

31. Cf. R. Dawkins, *The Blind Watchmaker* (1986). "Blindness" in evolutionary theory is also sometimes attributed to the "law-governed" nature of evolutionary events. Darwin was certain that the universe is governed by forces that can be described as "fixed laws," even if he could not

always discern or say what the laws are. But since he was certain that true laws identified true causes, and since forces (the ones Darwin was interested in) are not sentient beings, they obviously cannot see, and so when they "govern," they do so "blindly." This sort of account explains why the sorting processes of natural selection are sometimes called "blind."

32. J. Beatty (2010, 22–44) looks at recent experimental evidence that "random variation" is more important in speciation events than has been understood. A difference maker in these experiments is the discovery that the *sequence* of mutational changes, even within identical populations that exist within identical environments, occurs at different rates—even if the variations when they do occur are essentially the same. The fact that a variation in one group of organism happens *sooner* than in another will cause it to take an evolutionary path that may not be replicated and may even block other like mutations when they do appear from reproducing and spreading.

33. Similarly, when Dawkins denies that "natural selection is a theory of chance" (2006, 113–4, 120), he is not referring to the entire process of evolution, including variation, but only to the part of it that "selects" winners and losers in the struggle for life (i.e., natural selection). He is arguing in this section against "design" in the production of the natural order, not against chance variation as part of the production process.

2

Chance Transport

DARWIN ALWAYS BELIEVED that one of the great strengths of his theory is that it could explain large classes of facts. This virtue, he believed, more than offset what were regarded by the earliest critics of the *Origin* to be its two greatest deficiencies: no directly observed evidence by any human that species actually did transmutate into new species, and the absence either in the living or fossil record of organisms of the countless intermediate forms among different species that Darwin's theory would predict to have existed. Darwin responded to the former criticism by noting that species transformation is an imperceptibly slow process that would be unnoticed by human observers, even in principle. The arrival of a new species on earth is not an event anyone sees, only one that is recognized after it happens: new species are coronated as such nearly always post hoc (Dennett, 1995, 96 ff.). As for the latter criticism, Darwin consistently underscored the "imperfection of the geological record."

One class of facts that drew Darwin's attention from an early time was the idea of "chance transport." This seems to have been an important transitional idea for Darwin. It is hard to think about organisms able to move from one locale to another one far away by virtue of special equipment without thinking about where the special equipment came from. The new idea did obviate the need for a theory of special independent creation for many creatures, but it also brought forward a new question: from what causes do these myriad special abilities for travel arise? That question brought Darwin close to what would become his new line of thinking in the *Notebooks*: the question of how new variations arise, and how their appearance may play a role in evolution. After all, seemingly fortuitous transport leads directly to the question of the origin of the features that enable such transport to occur. And that, in turn, is a question about variations.

One may thus venture the opinion that "chance transport" may have been influential in leading Darwin to the discovery of his theory.[1] Certainly the *Journal of Researches*, first published in 1839, is replete with observations about the often puzzling ways organisms are distributed around the globe. Indeed, on the theory of "separate creation," geographical distribution often just did not make any sense. On the theory of descent through modification many of the puzzles disappeared.

In fact, geographical distribution is not just one fact needing explanation. It is, as Darwin observed, a "class of facts." In the class are the following (all noted in the *Notebooks CDN* B, C, D, and E, 1836–1839), posed here as questions:

(1) Why are large mammals found mainly on continents and not on islands? On the view of separate creation this distributional pattern would not be predicted (*CDN* B-81; B-115).
(2) Why are many organisms not particularly well-suited physiologically to their habitats (e.g., woodpeckers), so plainly designed to feed on insects living in tress, living in places devoid of trees? Why would a creative power decide to do that? (*CDN* B-115; B-130).
(3) Why do species habituating similar environmental stations that are separated by natural barriers (such as oceans, rivers, mountains) differ from one another to such a large extent? If species were separately created to suit their environments one would expect to see similarity or identity of organisms habituating similar stations (*CDN* B-158; B-280, n. 280-1).

(4) Why do organisms that greatly resemble each other, at least at the level of genera if not species, occupy regions of the earth that lie in proximity to each other rather than in like environments? For example, why do the fauna of the Galapagos more closely resemble species on the South American mainland, so very different in environmental characteristics, than species on other islands with environmental characteristics more like those of the Galapagos?

(5) Why are plants similar to their continental neighbors more likely to be found on islands than mammals? The theory of separate creation would presumably predict a more equal distribution (*CDN* B-192; B-194; B-221).

(6) Why is there greater similarity of flora and fauna at the same longitude than at the same latitude, even when current environmental conditions may closely resemble one another at the same latitude and may differ markedly at the same longitude? (*CDN* C-245).

(7) Why do differences in the organization of many animals differ in exact proportion to the length of time they have been physically separated from one another? In other words, the "time" of separation among organisms of the same class seems as important a factor in explaining similarities and differences as the geographical "distance" of separation (*CDN* B-224; D-23; E-23).

(8) Why do certain animals like mice, having a great facility for attaching themselves to "man," show up on remote islands whereas other animals, just as well suited to the island habitat, do not appear there? (*CDN* D-65).

Most of these questions are raised again in the 1842 "Sketch," the 1844 "Essay," and the *Origin*. The opening sentence of the chapter on "Geographical Distribution" in the *Origin*, for example, states:

> In considering the distribution of organic beings over the face of the globe, the first great fact which strikes us is, that neither the similarity nor the dissimilarity of the inhabitants of various regions can be accounted for by their climatal and other physical conditions (*Origin* [*Variorum*], 562: XI. 4).

This observation corresponds to items (3) and (4) above, from the *Notebooks*.

A "second great fact" about geographical distribution introduced by Darwin in the *Origin* concerns physical barriers to transport. Where barriers are absent, or might be presumed to have been absent in a distant past, it is found that organisms habituating adjacent areas in the economy of nature often closely resemble one another; whereas the presence of geographical barriers, either present or past, tends to correlate with the inhabitants on opposite sides of these barriers being very different from one another (*Origin* [*Variorum*], 563. XI: 12). This observation recapitulates item (3) identified above, from the *Notebooks*.

And so forth. Chapters XI–XII of the *Origin* reproduce, in the form of questions, more or less all of the observations and questions Darwin had already developed about geographical distribution in the *Notebooks* (1836–1839). The questions of latitude and longitude (563); of geological barriers (564); the similarity of creatures living in adjacent areas and the difference among creatures widely separated geographically, irrespective of environmental conditions (565–6); the relation between the time a genus has existed and the similarities of its various species (568); the presence of similar species of plants on islands to their continental cousins and simultaneous relative absence of similar species of mammals on those same islands (568–9) and others, are all treated. But by the time of the *Origin* Darwin had become bolder in his answers. The questions are now all answered by reference to the theory of descent through modification, with the important addition of confident assertions about how a theory of "chance transport" could account for the observed phenomena.

In brief, Darwin's theory was that most facts of geographical distribution could better be explained by transport than by separate creation. Put differently, the theory of separate centers of creation begged too many questions, all of which boil down to, "why this particular pattern of distribution?" To say, "the Creator so decided it" was for Darwin no answer at all. It amounted to saying that organisms are distributed as they are because that is how they are distributed. Darwin regarded such answers as non-answers, empty verbiage.[2] Besides, he believed, a better answer was available: the pattern of distribution could be explained by thinking about the possibilities and difficulties of transport of organisms from one particular center of first appearance to their eventual dispersal to various far-flung regions of the globe.

LYELL ON GEOGRAPHICAL DISTRIBUTION

Darwin did not invent the idea of geological transport, and he was frank in his acknowledgment. Suggestions that transport could account for at least some of the phenomena of distribution were present in several earlier writers. Among those Darwin acknowledged were von Buch, von Humboldt, Lamarck, and Étienne Geoffroy St. Hilaire (*CDN* D-69). Darwin did draw from these earlier observers. But he also believed that he made an important advance on them in terms of understanding nature's operations. While they had all made important observations, Darwin was the one who would "reduce" their "facts" into "laws." This, he claimed, was "the only merit" in his work "if merit there be" (*CDN* D-69).

But in fact the main source of Darwin's opinions that geological distribution could be explained by the idea of transport probably came from his friend and mentor Charles Lyell. Lyell's great book on geology, the *Principles of Geology*, accompanied Darwin on his *Beagle* voyage. He had the first volume (of what was to become a three-volume work) when he set out on his voyage. He was warned by Henslow, who knew he had a copy of the work (that Darwin received as a gift from Captain Fitzroy before the *Beagle* voyage began), not to believe everything it said, but Henslow nevertheless advised him to read it (Browne, 1995, 186–7). Darwin did read it, and with apparent care. The second volume came to Darwin while he was in South America, in 1832. This is the volume in which Lyell severely criticized Lamarck's theory. But it is also the volume in which Lyell devoted three chapters to what he called "Geographical Distribution" (*Principles*, vol. 2, chapters V–VII).[3]

In these three chapters Lyell made reference to "transport" to account for the geographical distribution of plants and animals on at least five occasions. He observed, for example, that the dispersion of fish ova across various geographical areas might best be explained by the assistance in transport provided by the water beetle, that seeds might be transported from here to there by birds, and that alligators might have transported organisms from one shore to another on the mud attaching to their claws (*Principles*, 5th ed., 1837, vol. 3, 134, 140, 159, 135). All of these observations may well have prompted Darwin to see the mystery of distribution to be solvable by reference to the idea that organisms are transported from one place to another by a variety of means.

But should it be said that such transport is "accidental" or "by chance"? That idea introduces a new level of complexity. It is one thing to say that the Creator endowed organisms with just the mechanisms He foresaw would be necessary for them to find their way from one geographical home to another, in response to changing environmental conditions. Darwin in fact did entertain this possibility in the early *Notebooks* (*CDN* B-100). But it is quite a different thing to say that organisms just happened to migrate, by "chance." Did Lyell say anything in his speculations that might have moved Darwin in the direction of believing this?

In fact he did. Most of Lyell's discussion of geographical distribution centers on what he calls the "means" of transport of various organisms. Chapter 5 is given over to the transport of plants, Chapter 6 of mammals, birds, and reptiles, and Chapter 7 of fish, sponges, shellfish and other aquatic animals. The details of the various means of transport need not detain us. An example will suffice to give the general line of his thinking. Plants, to use that example, are transported mainly by the dispersion of their seeds. Lyell observes, following previous analyses by several other travelers and naturalists, that seeds assume a variety of shapes and sizes that permit them to be carried by ocean currents, river streams, hurricanes, whirlwinds, birds, the fur of animals, and other "contrivances" to lands far distant from where they originate. Lyell also shows that the distribution of plants across the globe gives strong indication that most species of plants originated in one place and were thereafter transported by the above means to other places. Ocean currents, wind patterns, bird migrations, and so forth all give evidence that this is how plants disseminated. Lyell was implicitly arguing against a theory of "multiple centers of creation" in favor of a theory of single centers followed by migrations.

Thus, the idea of transport and its various mechanisms was not new for Darwin. It was not even new for Lyell. It had become an accepted way of thinking about organic distribution among many naturalists by the 1830s or earlier, even if it was not accepted by everyone. But Lyell's discussion of these matters implicitly raised two additional questions that would prove to be of special importance to Darwin. The first was, how did these dispersed species originate (in the single loci of production) in the first place? Lyell believed, when he wrote the *Principles*, that each species was originally a "separate creation," the product of a "creative force" whose offspring thereafter migrated.[4] Lyell did not abandon this conviction even after the appearance of the *Origin*. Allow as much dispersion by "natural means" as you will; the beginning of every species had to be sought in a separate "mysterious" act of creation. Darwin, of course, disagreed. Species descend through modification from other species.

More to our concern here is the second question that Lyell's discussion raised: are the various transportals of plants and animals "designed," or are they rather "accidental"? Lyell did in fact used the expression "accidental transport" (or equivalents) in several passages of the *Principles*. For example, he cites with favor "one of our ablest botanical writers, Keith" (i.e., P. Keith, *A System of Physiological Botany*, 1816, vol. 2, 405): "The mountain-stream or torrent washes down to the valley the seeds which may accidentally fall into it" (quoted in *Principles*, vol. 2, 76). Or again, "The number of plants found at any given time on an islet affords no test whatever of the extent to which it may have cooperated towards this end [of geographical distribution], since a variety of species may first thrive there and then perish, and be followed by other chance-comers like themselves" (vol. 2, 77). The idea that some transports are by chance is repeated yet again, in connection with the unlikely (but certainly actual) distribution of some seeds by their being transported inland from sea-sides by wading birds: "Let such an accident happen but once in a century, or a thousand years, it will be sufficient to spread many of the plants from one continent to another" (vol. 2, 81). These several expressions seem to indicate that Lyell was sympathetic to the view that transports can be and are often accidental.

That view receives additional support from other phenomena observed by Lyell, especially the complexity of interactions among organisms and their environments.

Such complexity, Lyell might have been suggesting, would make the idea of designed transport difficult to maintain. After all, such transportals, to be successful, required the cooperation of several, often many, simultaneous factors that, on casual inspection, seem to be unrelated on any theory of design but perfectly understandable on a theory of fortuity. Consider seeds that are transported in the fur of animals. For transport to be successful, it is necessary, first, that the seeds attaching to the fur are properly "hooked" to be able to hitch a ride; second, that the fur be of the right sort to take the hook; third, that the animal roves to new and distant geographical settings; fourth, that the hooks on the seeds enable them to become detached from the fur at some point; fifth, that the new environmental conditions are receptive to the transported seeds; sixth, that the seeds have the power to retain their vitality through long periods of time—weeks, months, perhaps even years; and so forth. That is a lot of "coincidence" to be able to sustain a serious case for "design."

Lyell seems to have understood the difficulty. In a memorable passage of the *Principles* he all but spelled it out. After discussing the preservation of seeds of some fruit-bearing plants, Lyell observes:

> The sudden deaths to which great numbers of frugivorous birds are annually exposed must not be omitted as auxiliary to the transportation of seeds to new habitations. When the sea retires from the shore, and leaves fruits and seeds on the beach, or in the mud of estuaries, it might, by the returning tide, wash them away again [or otherwise destroy them]; but when they are gathered by land birds which frequent the sea-side, or by waders and water-fowl, they are often borne inland... and may be left to grow far up from the sea.... Let us trace the operation of this cause [of seed preservation] in connection with others. A tempestuous wind bears the seeds of a plant many miles through the air, and then delivers them to the ocean; the oceanic current drifts them to a distant continent; by the fall of the tide they become the food of numerous birds, and one of these is seized by a hawk or an eagle, which, soaring across hill and dale to a place of retreat, leaves, after devouring its prey, the unpalatable seeds to spring up and flourish in a new soil (*Principles*, vol. 2, 81).

Consider for a moment the number of cooperating factors in dispersal: birds eating seeds of fruit-bearing plants, the same birds being destroyed by various unnamed natural causes, the seeds ingested by them having been left exposed on beaches or estuaries by retreating tides, other birds gathering up such seeds and bearing them far inland, or alternatively hawks and eagles devouring seed-bearing birds as prey and then leaving undigested seeds of birds thus devoured in new soil—can all of this be by design? That seems a "mighty stretch," as Darwin might have said.

Yet Lyell was sure all of this was by design. This becomes clear when he considered one final "means of transport," that which was effected by "man." What distinguishes transport effected by man from transport effected by natural causes is that the former is an "accidental" cause. By this Lyell meant that such transports were "unintentional" or "unconscious":

> But besides the plants used in agriculture, the number which have been naturalized by accident, or which man has spread unintentionally, is considerable (*Principles*, vol. 2, 82).

The words "accident" and "unintentional" are further elaborated some pages later:

> The most remarkable proof, says Decandolle, of the extent to which man is unconsciously the instrument of dispersing and naturalizing species, is found in the fact that [in many instances] the influence of man has surpassed that of all the other causes which tend to disseminate plants to remote districts. Although we are but slightly acquainted, as yet, with the extent of [man's] instrumentality in naturalizing species, yet the facts ascertained afford no small reason to suspect that the number which [man] introduces unintentionally exceeds all those transported by design (*Principles*, vol. 2, 84).

These passages tell us most of what we wish to know about Lyell's thinking about transport in general. Geographical distribution is to be understood as having been effected by two grand causes: transport of seeds, animals, and other organisms by natural causes that are "designed," presumably by the same creative power that brought species into being in the first place; and "accidental" transport of organisms of all kinds by "man." The latter are "accidental" only because they are "unintentional," or as Lyell elaborates in a later passage, "against [man's] will" (*Principles*, vol. 2, 84). Lyell may have picked up the latter idea from a comment he cites drawn from the voyages of Captain Cook whom, Lyell noted, had drawn attention to the "accidental transport" of organisms that attached themselves to canoes used by early human explorers (121). Even humans, Cook believed, had been transported long distances across the surface of the globe "unintentionally," through the vagaries of wind and ocean currents.

What Lyell does not address is a question that certainly interested Darwin: are these "accidental transports" effected by human voyaging themselves really "accidental?" Certainly they are accidental from the standpoint of human intentions: humans did not deliberately or willfully transport organisms from one location to another, at least in many cases (in other cases, of course, they did, in which instances the phrase "accidental transport" would be inappropriate). But is it possible that these accidental human transports were also "designed," in the sense that a higher creative power intended man to effect the transports, behind man's back, so to speak? To guess at Darwin's answer—and it cannot be much more than a guess—we shall need to turn to Darwin's reflections about the larger question of transport in general.

DARWIN ON GEOGRAPHICAL DISTRIBUTION

Darwin was attracted to the question of geographical distribution of organisms, as already noted, because of the light a proper understanding would shed on the question of the origin of species. His indebtedness to Lyell's prior reflections is apparent from what he wrote in the *Notebooks*. Not only did he remind himself to "remember Lyell's arguments" as he began his own early reflections about geographical distribution (*CDN* B-10), but he also borrowed many of Lyell's examples—floating seeds, wind and ocean currents, related forms of organisms in adjacent habitats, and so forth (*CDN* B-81, B-102). But did he get the idea of "chance transport" from Lyell?

Such clues as there are resolve themselves into two sorts: one, what did Darwin believe about the extent of the operation of the "creative force?"; and two, what did he have to say about the role of "accident" in transport? Both are important in discerning what Darwin believed, especially if they add up to the same conclusion. If the creative force is responsible through foresight and design for providing organisms with mechanisms of transport that would enable them to move to suitable environments around the globe as necessary, then accident would seem to play no role in transport. Conversely, if transport really is "accidental," in the sense of unplanned and undesigned, then doubts begin to be raised about the potency—and by implication the existence—of the creative force.

It is clear that almost from the beginning Darwin rejected a role for a directly intervening creative force to explain natural phenomena. For one thing, as he often said, to invoke a "creative power" to explain species origins, distribution, or anything else about species, was a non-answer. He observed early in *Notebook* B that such an "explanation" gives a convenient "hiding place for many [otherwise unintelligible] structures" (*CDN* B-99). This kind of thought was a common theme in the *Notebooks* (mid-1837; e.g., *CDN* B-104; B-196; B-216; C-200; C-209e; C-222e; D-72) and remained a common theme throughout Darwin's life (see especially Chapter XIV of the *Origin*, all editions).

But Darwin did entertain another notion, at least for a time, that the "creative power," if one existed, though not acting *directly* on organisms, might act *indirectly* on them. This indirect action would be through secondary causes, sometimes referred to by Darwin as "laws." On this view, an omniscient creator set up the world at the beginning in such a way that the world would be regulated by constant or "fixed" natural laws. This particular insight may well have come to Darwin from his understanding of Newton.[5] Heavenly bodies, argued Newton, course through their heavenly paths as they do, not because the creator presides at every instant of time over the paths they should follow, but because they are governed by physical laws that ensure a proper behavior for each heavenly body:

> Astronomers might formerly have said that God ordered, each planet to move in its particular destiny.—In the same manner God orders each animal created with certain form in certain country, but how much more simple, & sublime let attraction [of gravity] to act according to certain laws such are inevitable conseque[nce] let animal be created, then by the fixed laws of generation, such will be their successors (*CDN* B-101).

The idea that nature is regulated by constant laws does away with the need for an actively intervening deity in any of nature's operations, whether regulating the movement of planets or determining the creation of species. But it does not answer a prior question: whence the laws? One might believe that natural laws "just are," and leave it at that. Or, one might try to account for the origin of the laws. Darwin began his reflections on this question by adopting the "deistic" position that the laws were themselves designed "in the beginning" by a creative force:

> The question if creative power acted at Galapagos [and] it so acted that bi[r]ds with plumage <<&>> tone of voice partly American North and South.—(& **geographical <distribution> division are arbitrary & not permanent. This might be made very strong if we believe the Creator creates by any laws, which I think is shown by the very facts of the Zoological character of these islands.**) So permanent a breath [a directly intervening creative force?] cannot reside in space before island existed (*CDN* B-98).

The idea seems to be that the only plausible way to account for the presence of non-native species on recently formed islands (like the Galapagos) is that they have been transported there by natural laws rather than "breathed" into life by a creative force after the islands were formed.[6]

But no sooner had he said this than he began to express doubts that it could be so. Could a creative force, even an "omniscient" one, really have foreseen all the travel needs of species descended from earlier progenitors? The following entry comes but two pages after the one just quoted:

> How does it come wandering birds such as sandpipers, not new at Galapagos? Did the creative force know that <<these>> species could arrive? Did it only create those kinds not so likely to wander? Did it create two species closely allied to Mus[cicapi] Coronata, but not coronata? We know that domestic animals vary in countries, without any assignable reason (*CDN* B-100).

CHANCE TRANSPORT

To what extent did Darwin accept "accidental transport"? If transport is not accidental, it would seem to be the case that it is purposeful or designed. If, on the other hand, it is accidental, one might plausibly infer from this that neither creative power nor law exists to direct the wanderings of organisms to designed or necessary (law-governed) destinations.

It seems clear that Darwin understood these alternatives and grasped the implications for accepting one view rather than the other. If you accept special creation or necessary laws, no room is left for chance transport; and if you believe in chance transport, you seem to deny a place for a creative power or unfailing laws. As Darwin worked through this conundrum, he showed greater concern with the implications for the existence of a creator than of laws.

> The common mushroom & other cryptogamic plants same in Australia & Europe.—if creation be absolute theory, the creation must take place as when creator sees the means of transport fail.—otherwise no relation between means of transport & creation exists.—pooh [in other words, pure nonsense]. may have been created at many spots & since disseminated (*CDN* C-240e).

To this point Darwin takes his stand on the side of chance transport as opposed to special creation. "Transport," he urges elsewhere, is a much simpler explanation for geographical distribution than separate acts of creation. The theory of separate creation is no explanation, and besides leaves too many facts of distribution simply unexplainable (*CDN* D-115). Moreover, "we know many seeds might be transported, some blown, floating trees" (*CDN* C-100), and we know also that animal organisms of all kinds are equipped for transport in ways that would easily explain where they exist and do not exist: "Let the powers of transport be such, & so will be the forms of one country to another.—Let geological changes go at such a rate, so will be the number and distribution of species" (*CDN* B-102). "Species few in relation to difficulty of transport. For instance the temperate parts of Teneriffe, the proportion of genera 1:1" (*CDN* B-158). "Well to insist upon large Mammalia not being found on all is[lands]; (if [mammals are] act of fresh creation why not produced on New Zealand?)" (*CDN* B-115). Even the geological distribution of "man" seems perfectly explainable by reference to man's powers of transport (*CDN* D-65).

With doubt thus cast on an intervening deity Darwin was free to give greater room to chance transport. Lyell had, as observed earlier, already by 1833 referred explicitly to the "accidental drifting" of canoes carrying men "unintentionally" to new, previously inhabited regions of the globe, and Darwin cited Lyell's observation with approval at *CDN* E-65 (cf. *CDN* E-65, n. 1). But he had much earlier come to similar conclusions about the transport of other organisms. For example, in referring to a conversation Darwin had with Henslow in May 1838, Darwin observed:

> Henslow, in talking of so many families on Keeling, seemed to consider it owing to each, being fitted [by the creator?] for transport. May it not be explained by mere chance? (*CDN* C-100).

Similarly, in reference to the unusual migration patterns of certain birds, Darwin speculated:

> If the line or bands of country...gradually separated the birds might yet remember which way to fly. There is a kind of Wren (Bebyk??) which seems common in Rocky Mountains & on one lofty isolated spot on the Alleghanies [*sic*] to which it migrates every year; and probably a <<chance>> wanderer like the first pair of Pipra flycatcher (*CDN* C-255).

The insertion of "chance" after the original composition of this entry (signified by the << >>) suggests that Darwin was not just carelessly tossing around a handy expression, but rather was making a deliberate decision to use a word that gave an accurate reflection of his thinking.

Darwin did not pretend to know all the details of plant and animal transport, and freely confessed his ignorance from time to time (e.g., *CDN* B-224). When the means of transport are unknown in particular cases, it is nevertheless "more philosophical" to state "we do not know" than to revert to the theory of special creation (*CDN* B-194). "We do not know," in fact, allows an opening for a return to "laws" that Darwin would later exploit. "Chance" need not mean, and for Darwin did not mean, "no cause," or "no law." It means, rather, that whatever the cause or law, we should assume nature to proceed in accordance with laws, and until we discover what they are we may just as well say, "laws unknown." In view of the absence of complete knowledge, "chance" will often be found to be a sufficient explanation for the migration of species across the globe. In a particularly fulsome and revealing passage in *Notebook* B Darwin had this to say:

> The number of genera on islands & on Arctic shores evidently due to the chance of some ones of the different orders being able to survive or chance having transported them to new station.—When the new island splits & grows larger species are formed of those genera & hence by same chance few representative species. This must happen, and then enquiry will explain representative Systems. Of this we see example in English & Irish Hare,—Galapagos shrews & when big continent many species belonging to its own genera (*CDN* B-221).

GEOGRAPHICAL DISTRIBUTION IN THE 1842 "SKETCH" AND 1844 "ESSAY"

The phrase "chance transport" is mentioned three times in this very early *Notebook*, B. In the 1842 "Sketch" Darwin reaffirmed the inadequacy of the theory of separate creations as an explanation for geographical distribution (1842, 1971, 70). He also reaffirmed the operation of the Creator, if one exists, through the agency of "intermediate means," to be understood as "fixed laws" (45–6; 59, where the analogy to the motion of planets in the field of astronomy is again invoked). Such a view, Darwin now maintained, was a "more exalted view" of the Creator than the view that brought a divine Creator in at every stage of organic creation. And finally, Darwin insisted on "accidental" or "chance" transport to account for the presence (or absence) of organisms in various regions of the world (44; 68–9).

The much longer 1844 "Essay," a work that Darwin understood *might* be brought forward in published form (if, for example, he were suddenly to die, in which event his wife Emma was instructed by him in writing to find an editor and a publisher for the work), continued with many of the same arguments. Affirmed again are the points: (1) that the Creator, if one exists, works through "secondary means," (i.e., laws) (116; 154; 194; 253–4), certainly not through separate acts of creation (132; 186). Many of the same examples of the means of transport are invoked—birds and animals ingesting seeds and transporting them vast distances from point of origin, water beetles carrying fish ova from place to place, wind and ocean currents carrying fish and birds long distances, and so forth (183). And again Darwin invoked "accident" and "chance" as explanations for how transport occurs. By this time he was clearer than before that "chance transport" really means "we do not know how this occurs," but at least we know that the placement of organisms is not to be explained by the theory of separate creations. It was for Darwin "more philosophical" to simply admit ignorance about how organisms are transported than to fall back on the non-explanation that distribution is just a matter of the Creator placing organisms here and there because He chose to do so. Darwin had made this same point much earlier in the *Notebooks* (*CDN* B-188; B-194).

In the 1844 "Essay," however, we are introduced to a new word in Darwin's lexicon that started to replace the words "chance" and "accident": "occasional transport" (114–5; 196–7; 253–4). The introduction of this word begins a trend in Darwin's writing, one that can be summarized by saying that "chance" and "accident" had to give way to a different mode of expression, one that seemed to replace "fortuity" by the more neutral idea of "once in a while." In the 1844 "Essay" this substitution of the new term is unannounced, and so nearly invisible to the casual reader. But "occasional transport," though equivalent in meaning to "accidental transport," has the effect of removing one strong objection that Darwin could anticipate his theory would encounter—a chance-governed world. We shall pick up this thread when we turn to consider Darwin's views in the *Origin*. But first we must take a brief detour through the correspondence, 1855–1866.

DARWIN'S DISPUTE WITH HOOKER

By 1855 the terms of debate about geographical distribution had shifted for Darwin. He no longer saw any need to argue with separate creationists, even though many of them were still around arguing for that position. Darwin simply turned his back on their claims. He had already said enough in his earlier writings to dismiss the value of that theory. More importantly, he now had a new antagonist, one more worthy of scientific concern. This was the geologist E. Forbes, then supported by the entomologist T. V. Wollaston, who promoted the theory that geographical distribution could best be explained by land migrations of various species over previously connected land masses, now separated by vast ocean expanses. This is the so-called "Atlantis" theory. According to this view, most of the territory occupied by continental Europe and the northern islands (England, Ireland, and others) was previously one vast connected land mass. At some point in a remote geological past these connected land masses had been sundered by natural geological processes that had submerged parts of this connecting land, such that the remaining parts were separated by bodies of water. Because "natural barriers," like oceans, would presumably inhibit migration of species from one point to another, the true explanation for similarities of species on surviving land masses separated by large bodies of water must be that migrations had occurred across land before the natural barriers came into existence. This was still a theory of "transport," not separate creation, but it was one that did away with the need for any special contrivances in organisms that would have been necessary to carry them across large bodies of water.

Darwin immediately rejected this theory. His main reason was that it involved too many hypotheticals:

> I have had some correspondence with W[ollaston] on this & other subjects, & I find he coolly assumes (1) that formerly insects possessed greater migratory powers than now (2) that the old land was *specially* rich in centres of creation (3) that the uniting land was destroyed before the special creations had time to diffuse, & (4) that the land was broken down before certain families & genera had time to reach from Europe or Africa the points of land in question.—Are not these a jolly lot of assumptions? (*Correspondence of Charles Darwin*, hereafter *CCD* [March 7, 1855], Darwin to J. D. Hooker).

The theory disgusted him: "my blood gets hot with passion & runs cold alternately," he wrote to Hooker, "when geologists so casually invent no-longer existing land-masses to explain annoying facts of distribution" (*CCD* [June 16, 1856], Darwin to J. D. Hooker). Much easier to believe, Darwin thought, that the migrations had occurred over bodies of water between landmasses that had long, if not forever, been separated. But his theory would have to be corroborated by convincing evidence.

The dispute between Forbes/Wollaston and Darwin was enacted in real time between Darwin and his closest collaborator and strongest supporter, J. D. Hooker, in the 1860s. Hooker was at first a strong skeptic about Darwin's view. He thought the theory of previous land bridges was plausible, and he also thought that "transport" across ocean waters would not work to explain many instances of plants (he was a botanist) showing up on widely separated continents and islands. A tangled controversy ensued between the two men. Darwin appealed to "transport," whereas Hooker defended connected landmasses. Darwin was puzzled by Hooker's stance:

> I cannot make exactly out why you w^{d}. prefer continental transmission, as I think you do, to carriage by sea: with your general views, I sh^{d}. have thought you w^{d}. have been pleased at as many means of transmission as possible—For my own pet theoretical notions, it is quite indifferent whether they are transmitted by sea or land, as long as some, tolerably probable way is shown. But it shocks my philosophy to create land, without some other & independent

> evidence. Whenever we meet, by a very few words I sh[d]. I think more clearly understand your views (*CCD* [6 June 1855], Darwin to J. D. Hooker).

The controversy provoked Darwin to undertake numerous experiments on the ability of seeds to withstand ocean travel without losing reproductive potency. (Darwin had much earlier experimented on the responses of different seeds to different soil conditions, but the "Atlantis" theory seems to have driven him to the saltwater experiments.)

Darwin's private exchanges with Hooker about this issue continued to broil through a dozen years, from 1855 to 1867. It reached two pitches of climax, the first in 1862 centering on Hooker's views about the flora of Greenland, all of which suggested that transport could not be an explanation for the distribution on that island of the various species. Darwin allowed that Greenland was a problem for his theory and that Hooker *might* be right *in this case*.[7] The second occurred in 1866. The particular occasion of this dispute seems to have been the unusual distribution of species on the Azores, connected in so many respects with other islands, but so different in the species of plants found on them. Darwin continued to hold out for transport, Hooker (at first) for something else, he knew not what.[8] By August 1866 Hooker was perplexed, telling Darwin that *neither* theory could account for the phenomena, not transport, and not former connected landmasses. This confession by Hooker gave Darwin his opening:

> We both give up creation & therefore have to account for the inhabitants of islands either by continental extensions or by occasional transport; now all that I maintain is that of these two alternatives, one of which must be admitted notwithstanding very many difficulties, that occasional transport is by far the most *probable*.
>
> I go thus far further that I maintain, knowing what we do, that it w[d]. be inexplicable if *unstocked* isl[ds] were not stocked to certain extent at least, by these occasional means (*CCD* [August 8, 1866], to J. D. Hooker).

Under this onslaught of unwavering argument and previous evidence from Darwin, Hooker finally began to capitulate to Darwin's view. In an address before the British Association for the Advancement of Science at Nottingham in August 1866, Hooker finally allowed that the transport theory might have merit at least equal to the "continental extension," or "former land-bridge" theory of Forbes and Wollaston. He did not proclaim solidly on either side, but he did give Darwin's theory equal time. Darwin was gratified. The transport theory had at least found a sponsor of scientific respectability and objectivity. The audience responded sympathetically to Hooker's address, and Darwin's transport theory was thus thrust into the forefront of probable theories for explaining geographical distribution.

Origin of Species

But Darwin had already gone public with his views about transport well before his private exchanges with Hooker on the subject—in the *Origin* in 1859. In many respects, Darwin's thinking about geographical distribution did not change from where it was already in the *Notebooks*. We again find: (1) the rejection of the theory of separate creations (*Origin* [*Variorum*], 571: XI. 73); (2) the belief that geographical distribution, like other natural phenomena, is governed by "universal laws," the laws having to do with the various means of transport (570: XI. 69); (3) confession of ignorance about what some or many of those laws governing means of transport are (568: XI. 49–51; 607: XI. 274); (4) an enumeration of many of the known or suspected means of transport, including seeds transported by former glacial ice-floes, seeds floating in ocean currents, seeds hitching a ride in the wool of animals, and seeds being carried in the guts of mammals and birds (573–83: XI. 81–155); (5) the importance of barriers—bodies of water, mountain ranges, and so forth—in preventing transports that otherwise should be expected to occur (563–67: XI. 12–40); and (6) the "inadmissibility" of Forbes' "former land-bridge" theory (573–4: XI. 87–91).

But in the *Origin* we find an important new development. Whereas previously Darwin had spoken of "accidental transport" without hesitation or qualification, we now get in the *Origin* some backtracking. For one thing, whereas in the *Notebooks* transport was always by "chance" or "accident," in the *Origin* it was increasingly "occasional." We saw this substitution make its first appearance—but only irregularly—in the 1844 "Essay." In the *Origin* it has become the typical expression, and quite consciously and deliberately:

> I must now say a few words on what are called accidental means [of geographical distribution], but which more properly might [should: 5th edition] be called occasional means of distribution. *Origin* [*Variorum*] 575: XI. 99 [and 99*e*]; the sentence is followed by Darwin's report of the results of various experiments that he had conducted on transport).[9]

"Accidental" has now become "occasional," a word that really does not mean "accidental" at all. And we are told not just that "occasional" is a suitable substitute for "accidental," but that it is a substitution we "should" make. In other words, we *should* get rid of the word, if not idea, of "accident" in the processes of distribution.

What accounts for this shift in usage/meaning by Darwin? The obvious answer is that Darwin had come to see that law-governed processes cannot be "accidental," by definition. That, in fact, seems to be his explanation:

> These means of transport are sometimes called accidental, but this is not strictly correct: the currents of the sea are not accidental nor is the direction of prevalent gales of wind (*Origin* [*Variorum*], 582: XI. 145).

The apparent inference is that geological and meteorological processes are law-governed, and so not accidental, and so, therefore, neither is transport:

> A few facts seem to favor the possibility of their [fresh-water fish] occasional transport by accidental means [changed in fifth edition to: "it is probable that they are occasionally transported by what may be called accidental means" (*Origin* [*Variorum*], 612: XII. 12; 12.x-y:*f*)].

The means of transport "may be called" accidental, but in fact the transport is better called "occasional." The idea has shifted from fortuity in geographical distribution to periodicity. Ocean currents, winds, and other such processes *may* transport organisms, but again they may not. They can be expected to do so only from time to time, not invariably. It is because of the variability of the processes that it is better to say they operate "occasionally" than that they operate "accidentally." So seems to be Darwin's intention when he made the substitution in words.

It should be noticed that Darwin's employment of the phrase "occasional" transport, to this point, only makes reference to natural *processes*, wind and ocean currents, climate changes, and others. It does not refer to organic *structures* that enable organisms to exploit these natural processes. Yet processes cannot work to transport organisms unless organisms are properly equipped with "contrivances" that enable them to be transported. Such contrivances would include the ability of some seeds to withstand lengthy periods of saltwater immersion, some seeds being equipped with hooks for attachment to bodies of roaming animals, some fish ova being able to endure vitality even during ingestion and digestion by water beetles, and so forth. Successful transport depends on both—suitable natural processes and suitable organic structures. Is it possible that Darwin's notion that transport is often "accidental" led him to wonder whether structures and not just processes are accidental?

This question takes us directly to Darwin's speculations about organic variation and its causes. As we shall presently see, Darwin was fond of referring to variations as often being due to "chance," at least through the first edition of the *Origin*. It seems a plausible hypothesis that he was drawn to this expression from his prior reflections on transport. How can one think about the processes of transport without thinking about organic structures that enabled transport—the ability of some seeds to withstand immersion in sea water or enlodgment within animal guts without losing their ability to germinate, seeds being equipped with hooks for fastening onto animal fur, or conversely, that disabled transport—the wingless character of some birds, the fused elytra of Madeira beetles, and so forth. Structures cannot be ignored when weighing the various issues that geographical distribution raises.

The idea of a connection between processes and structures was never clearly spelled out by Darwin. Perhaps he did not even really notice it. But from his *Notebooks* to the *Origin* we find expressions in Darwin's writing that show he often had structures in mind as much as processes. For example, the distribution of flora found on islands and arctic shores is as much due to "chance having transported them" as to "the chance of some . . . being able to survive [such transport]" (*CDN* B-221). Other patterns of distribution depend on "the time of separation" and the "facility of transport" (*CDN* D-23). The distribution of mice on islands is surely explained by "the facility with which they attach themselves to man" (*CDN* D-65). In reference to seeds being carried long distances by ocean currents Darwin observed: "When I show that islands would have no plants were it not for seeds being floated about, I must state that the mechanism by which seeds are adapted for long transportation, *seems* to imply knowledge of whole world—if so doubtless part of system of great harmony" (*CDN* D-74e).

In these passages and others Darwin is concerned with structures, not just processes. But in that case another question arises: how did these structures that so well suited organisms for transport, or just as assuredly prevented them from being transported, arise? One cannot say only that "accidental" or even "occasional" processes and environmental conditions caused transport to occur. After all, earlier generations of related species (progenitors of the travelers) did not get transported, even when conditions and processes were no doubt the same or similar. Variants had to arise that possessed the features necessary to take advantage of the means of transport at their disposal. It is a conjecture, but I believe a plausible one, that Darwin arrived at his important idea that variations are "by chance" through his reflections on transport and his early belief that the best way to account for geographical distribution is that it is purely fortuitous: some transports did happen to happen, others happened not to have happened, and between the two no discernable cause can be assigned.[10]

Our only remaining question here is why Darwin started to use the language of "occasional" in the place of "accidental" by the time he got to later editions of the *Origin*. We can only guess. As noted, with respect to natural *processes*, as contrasted with *structures*, Darwin claimed in the *Origin* that these processes are not accidental—by which he evidently meant they were law-governed. But he had come to think of all natural processes as law-governed much earlier, as early as 1837. Yet he did not come to change his wording in published writing until 1859, and especially 1868 (the fifth edition of the *Origin*). It seems therefore likely that his change in wording was brought about less by a change of heart about the facts of nature than a change of mind about how best to present his theory. In particular, the *words* chance and accident were red flags to his audience. Darwin saw that he needed to get rid of them. He did the same when he spoke about the causes of variation: at first they are often "by chance," and later, "chance [as a cause of variation] is a wholly incorrect expression."

NOTES

1. Already in an early passage in *Notebook* B (late 1837) Darwin made this remarkable statement: "Let the powers of transportal be such and so will be the form of one country to another.—Let geological changes go at such a rate, so will be the numbers and distribution of species!!" (*CDN* B-102). Later in life, after the *Origin* had been published, Darwin recalled to his friend Hugh Falconer, "Let me explain how it arose that I laid so much stress on Natural Selection, and I still think justly: I came to think from *Geographical Distribution* &c. &c. that species probably change; but for years I was stopped dead by my utter incapability of seeing how every part of each creature (a wood-pecker or swallow for instance) had become adapted to its conditions of life. This seemed to me and does still seem the problem to solve, and I think natural selection solves it, as artificial selection solves the adaptation of domestic races for man's use" (*CCD* vol. 10, 440 [October 1, 1862], to H. Falconer, emphasis supplied).

2. "We have absolute knowledge that species die & others replace them—two hypotheses fresh creation is mere assumption, it explains nothing further, points gained if any facts are connected" (*CDN* B-104).

3. These three chapters were retained by Lyell through the ten editions of the *Principles*, essentially unaltered. Darwin's later references to these chapters seem to be based on the fifth edition, not the first (*CDN* B-115 and n. 115–1 where the editors refer to Darwin's extensive marginalia in the chapters on geographical distribution in Lyell's fifth edition [1837, 27–76]). Page references are to the fifth edition of the *Principles*.

4. *Principles of Geology*, 5th ed., vol. 1, 161; vol. 2, 23–4. A subsidiary question is whether Lyell believed in an original single locus of creation, from which all existing organisms radiated, or in multiple loci. Darwin considered this question in *CDN* B-155, deciding in favor of multiple loci; cf. *CDN* B-155 and n. 155–1.

5. Other naturalists prior to Darwin had argued for much the same view, notably F. Cuvier and Étienne Geoffroy St. Hilaire. Darwin acknowledged their views on this subject, without fully endorsing them, in the B *Notebook*: cf. *CDN* B-111–15.

6. Darwin again famously referred to life being "breathed into" living creatures in the final sentence of the *Origin*, which is often quoted. In the same sentence he compared the continuous production of new species through time to the continuous "cycling of planets according to the

fixed law of gravity," implying at best that the creative force, if there is one, has left the governance of the earth to the operation of fixed laws. Somewhat strangely, Darwin changed this last sentence of the *Origin* in the second edition to read "breathed by the Creator into a few forms or into one" (*Origin* [*Variorum*], 759: XIV. 270). He later regretted making this change as an unintended capitulation to the "Pentateuchal" point of view.

7. The Greenland controversy between Darwin and Hooker occupied their concentrated attention in their correspondence between February and March 1862: cf. *CCD* [February 25, 1862] Darwin to Hooker; [March 3, 1862] from Hooker to Darwin; [March 7, 1862] from Darwin to Hooker. Darwin brought Asa Gray into the fray in July 1862 ([July 28, 1862] from Darwin to Asa Gray).

8. See the flurry of letters between Darwin and Hooker in July–August 1866: *CCD* [July 30, 1866] Darwin to Hooker; [August 1866] Darwin to Hooker; [August 4, 1866] Hooker to Darwin; [August 5, 1866] Darwin to Hooker; [August 7, 1866] Hooker to Darwin; [August 8, 1866] Darwin to Hooker; [August 9, 1866] Hooker to Darwin.

9. The term "occasional" as an apparent substitute for "accidental" had already made an appearance in the *Origin* in Chapter IV, "Natural Selection" ([*Variorum*] IV: 190, 160–68, all editions), in connection with intercrossing: "As yet I have not found a single case of a terrestrial animal which fertilizes itself. We can understand this remarkable fact, which offers so strong a contrast with terrestrial plants, on the view of an occasional cross being indispensable, by considering the medium in which terrestrial animals live, and the nature of the fertilizing element; for we know of no means, analogous to the action of insects and of the wind in the case of plants, by which an occasional cross could be effected with terrestrial animals without the concurrence of two individuals." Darwin's reference to insects and wind in helping intercrossing among plants is suggestive of the idea of "chance transport," supporting the conjecture that chance transport and chance intercrossing were connected ideas for Darwin, both, however, now being consistently rendered by the very different expression "occasional." "Occasional crossing" (like "occasional transport") had already appeared in the 1844 "Essay" (*EBNS*, 102).

10. Kohn and Hodge (in Kohn, 1985, 185–204) study "chance variation" in the *Notebooks* but do not give much attention to "chance transport" as a separate concept. Nor do they appear to notice "chance transport" as a potentially important transitional idea for Darwin.

3

Causes and Laws of Variations

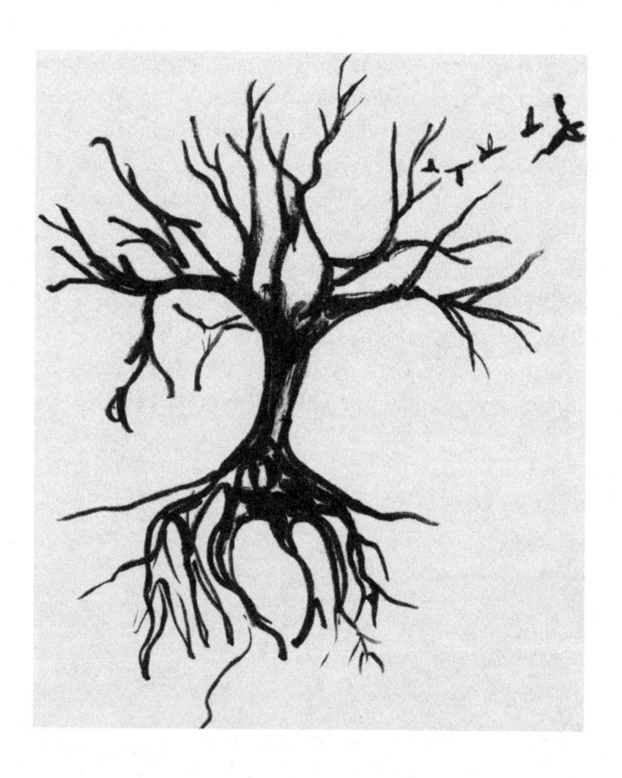

FROM HIS REFLECTIONS about biogeographical distribution and how transport might contribute to a solution to puzzles raised by it, Darwin (as noted in chapter 2 of this volume) may well have been led to a further question: how do the "contrivances" that make transport possible arise? For Darwin that was a question about "causes of variation."

The subject, he knew, was difficult and complicated, but he did not shy away from it. He knew he had to confront it. A great deal of his speculative efforts in the first five chapters of the *Origin,* not to mention earlier reflections going back at least to 1838, was dedicated to solving it. This chapter traces his thoughts on this question.

When natural selection "acts" to select organisms for preservation or destruction, it can do so only on organic matter that is already given, the "raw material" of selection, as it is often called. The raw material is organisms. All organisms vary slightly from generation to generation, but not all variations play a role in evolution. If current equipment, habits, and structures are well suited to current environmental conditions, and neither the former nor the latter change much, then one would expect existing species to continue on more-or-less unchanged.[1] If, on the other hand, either environmental conditions or structures and habits were to change, new possibilities arise. If the new structures and habits are ill-suited to existing conditions, the organisms possessing them might well perish in the struggle for life. If conditions change without corresponding changes in structure and habits, the same fate, extinction, may result. But if organisms change in such a way as better to adapt them either to existing or changing conditions, the conditions are ripe for evolution.

The last case is the one that held the greatest interest for Darwin. We see again that the question has two parts: how/when/why do environmental conditions change? And how/when/why do organic beings vary over the course of time? The former question Darwin left to the geologists, mainly Charles Lyell, and he was inclined to accept their views without much question.[2] The latter question sent Darwin on a search for the causes of variation, and in his exploration of this he was a true pioneer.

LAWS AND CAUSES: HERSCHEL AND WHEWELL

Every edition of the *Origin* includes a chapter called "Laws of Variation" (Chapter V). There is some evidence that Darwin regarded this chapter as the centerpiece of the entire book, perhaps the centerpiece of all of his theoretical work on the species question.[3] Variation is the lynchpin of organic evolution. He believed no evolution can occur without it. And variation must come first. Natural selection is powerless without organic material to work on, and the sort of material it does work on is variations. The variations given up by nature determine the direction and pace of evolutionary change, so much so that one might almost call Darwin's theory a theory of "variational determinism."

Variation is the first step in the modification of species. New species cannot form unless members of existing species undergo variation. Darwin in fact sometimes referred to variations as "incipient species" (e.g., *Origin* [*Variorum*] 136: II. 69). Beyond that, when Darwin began writing his "big species book" in the mid-1850s, he referred to it informally as "my book on variation," not "my species book."[4] This work was postponed by the arrival, in 1858, of Wallace's manuscript and Darwin's realization that he would have to scrap (for the time) the "variation book" and undertake a shorter "abstract" that became the *Origin of Species*. But the seeds of the *Origin* were already present in the "book on variation," and some of the fruit made its way both into the *Origin* (not only Chapter V, but in the title of the *Origin* itself, where the word "variations" is replaced by the term "races") and into Darwin's later opus *Variation of Animals and Plants Under Domestication* (1868). (The rest of the original "big species book" was published only in 1975 under the title *Natural Selection,* edited by R. Stauffer.)

Darwin believed from an early time that variation, like all other natural processes, is law-governed, and also that all variations are "caused." Before we examine what he thought those laws and causes are, we must detour through a brief recounting of what he took "laws" and "causes" to mean. In particular, a distinction must be made between the two as they are used in Darwin's theory (Schweber, 1985, 48, in Kohn, ed., 1985). While related, they are not the same. The distinction comes out well in John Herschel's *Preliminary Discourse on the Study of Natural Philosophy* (1831) and William Whewell's *Philosophy of the Inductive Sciences* (1840). Darwin's understanding was heavily influenced by these two men (Darwin, 1958; Ruse, 1975; Thagard, 1977; Ruse, 1978; Gildenhuys, 2004; Richards, 1997), so it will be helpful to take a look at what they had to say about causes and laws before we turn to Darwin's views.

For both men, "explanation" of natural phenomena is the principal aim of science. And for both men this generally was a matter of finding "causes," especially, in the case of Herschel, "true causes" or what he often called *verae causae* (198).[5] A "true cause" furnishes a "complete explanation of the facts" under consideration (158); or, conversely, "the sole and sufficient cause of all the observed phenomena" constitutes an explanation (156). If the explanation is not complete, the search for additional causes is then necessary. If a "residual phenomenon" remains (i.e., one not yet explained by causes already identified), the search for additional causes becomes necessary until one arrives at the cause that explains all the phenomena under investigation. When one has found that cause, one has found a "true cause" and one thus has an "explanation" of the phenomena.

In addition, a cause is to be understood as a "force" or a "power" or an "efficacy" that has a "real operation" in nature (Whewell, 162; Herschel, 198). This means that when two phenomena are always connected in time and space (e.g., lightning and thunder), the first one is just that which explains the existence of the other. But, in the example, at least for Whewell, lightning is *not* the "cause" of thunder. Lightning is lightning. The "cause" of the thunder is, one might say, a third thing that is "apprehended by the constant activity of the mind" by which the cause is understood to connect the two phenomena as a matter of necessity. The "mind" apprehends the "necessary truth" of the cause's existence and identity (169). Cause is, in Whewell's expression, a "Fundamental Idea," an innate concept in the human mind. But, the mind does not just "make up" the cause, it "apprehends" it (also Herschel, 144).[6] Minds are structured so as to enable them to make such apprehensions, and if they were not structured in that way they would not be able to "make sense" of sensible perceptions. The "cause" has a real identity independent of the mind. The aim of science, thus, is precisely to "apprehend" these invisible causes.

Herschel was more concerned with "laws" than with causes. Had he known about Whewell's views (Herschel's book preceded Whewell's by nine years) he may well have found Whewell's account of causes to be too metaphysical or Kantian (Ruse, 1975), although the two men did agree that "causes" are "forces" that bring about phenomena. And indeed, like Kant, Whewell was arguing against the Scottish skeptical philosopher David Hume. Hume had denied any "real existence" of causes. Lightning and thunder are "real" because they may be perceived by human senses (in this case, sight and hearing, respectively). Causes are not real because they cannot be perceived by sense. Hume denied the existence of anything that cannot be directly perceived. Causes, thus, are figments or fancies of human imagination. Instead of saying, therefore, "x causes y," wherein "cause" is a "real agency or force," Hume declared that what is really meant by such expressions is simply "x always precedes y in space and time, as far as we know." The "as far as we know" is important because it keeps open new possibilities in light of new experience (Herschel and Whewell agreed with Hume on this qualification). But Hume's real novelty was his denial that causes are "real" forces or anything else "real." Herschel and Whewell, despite other disagreements, agreed with each other against Hume on this point. Darwin, as we shall see, sided with Whewell and Herschel: causes are "real forces" and a major aim of science is to apprehend them.

Laws, as mentioned, are different from causes. The distinction is best brought out in a memorable passage from Herschel's work:

> Finally, we have to observe, that the detection of a *possible* cause, by the comparison of assembled cases, *must* lead to one of two things: either, 1st, the detection of a real cause, and of its manner of acting, so as to furnish a complete explanation of the facts; or, 2dly, The establishment of an abstract law of nature, pointing out two phenomena of a general kind as invariably connected; and asserting, that where one is, there the other will always be found. Such invariable connection is itself a phenomenon of a higher order than any particular fact; and when many such are discovered, we may again proceed to classify, combine, and examine them, with a view to the detection of *their* causes, or the discovery of still more general laws, and so on without end (158–9).

On this account, in the investigation of nature, one is on the lookout for one of two things: "real causes" which "act" to bring things about; and "abstract laws" that are "assertions" about "invariable connections" among observed phenomena.

Unlike causes, which are "forces," laws are "assertions" about "invariable connections." Laws do not "act," in the same way that propositions do not "act."[7] Causes may of course be *expressed in* propositions, and to be useful to scientific understanding they must be. But they are *not* in themselves propositions. They are "forces." Laws are, by contrast, *only* propositions, not "forces." In addition, such propositions as are "laws" may involve no mention of "causes" at all (although they may well do so!). The only requirement for a law to be a law, at least in Herschel's view, is that it be a proposition asserting "invariable connections."

The distinction is subtle and is easy to overlook (as Darwin did). One often speaks of a "law of gravity" and other "natural laws," and such expressions may easily be taken to mean that "the law of gravity" *governs* the behavior of heavenly and earthly bodies (Herschel's favorite example: 198). But laws do not "govern," they only assert. In the example of gravity, for instance, "gravity" is a "cause" that "governs"; the "law of gravity" is an assertion about what this cause is and about how it works.

CAUSES AND LAWS: DARWIN

Darwin accepted Whewell and Herschel's understanding that causes are "forces" that bring things about in nature, although he avoided any metaphysical entanglements regarding Whewell's adjudication between Kant and Hume.[8] It is likely that he was more in the camp of Herschel on this question. But it is not clear to me that he grasped Herschel's distinction between "causes as forces" and "laws as propositions." He often spoke as though they are the same thing (e.g., *CDN* E-150; and below). At other times, however, he did employ the distinction but without acknowledging that he understood that he was doing so. It is his usage rather than any explicit statements he made that shows that this was so. In Chapter V of the *Origin*, for example, we find that not all laws are causes, but all causes can be stated as laws. Laws, on this view, encompass causes but also much more. A *cause* acts in nature to bring things about; a *law* gives only a correlation between or among phenomena, "a relationship" as Darwin called it in 1838 (*CDN* C-158), without pretending to say *why* the phenomena in question are correlated.[9]

One observes Darwin's evident confusion between "laws" (meaning "statements about correlations") and "causes" (meaning "forces" that "produce effects") even in the first paragraph of Chapter V. While the chapter is entitled "Laws of Variation," Darwin quickly came to assert that his real interest was in finding "causes": "[we are] plainly ignorant [in many cases] of the cause of each particular variation" (*Origin* [*Variorum*],

275: V. 5). The sentiment was repeated at the end of the chapter: "Whatever cause may be of each slight difference in the offspring from their parents—and a cause for each must exist—it is the steady accumulation, through natural selection of such differences [that] gives rise to [new species]" (320: V. 305). And in the passages in between Darwin often employed terms that suggest causes, not laws: variations are sometimes "due to" changing physical conditions, variations can be "attributed to" various factors, some changes in physical conditions have a "direct effect" on the structure of organisms, sometimes physical conditions "act indirectly" to produce variations, and so forth. The words in quotes denote causes. The search for causes of variation was Darwin's central concern in Chapter V, more than was his search for "laws."

Yet it would be incorrect to say that Darwin had no interest in laws. In fact, as we shall see, when Darwin finally came around to summing up his reflections in Chapter V on variations and how they occur, he gave both "causes" and "laws." It is not clear that he saw any difference. He began his summary statement of Chapter V by saying:

> Our ignorance of the laws of variation is profound. Not in one case out of a hundred can we pretend to assign any reason why this or that part differs, more or less, from the same part in the parent (*Origin* [*Variorum*] 317: V. 285–7).

Well, which is it: was Darwin looking for causes or for laws? "Reasons" connotes "causes," yet the statement is about "laws." His catalogue of factors immediately following the sentence just quoted contains both, and by commingling them Darwin seems to have been unaware that his discussion was in some measure confusing apples and oranges.

Even in the first edition of the *Origin*, Darwin was able to identify 14 "laws" of variation, but not all are properly called "causes." To make our way through his complicated discussion it will be useful to discriminate (as Darwin did not do) among three different kinds of explanation for variation: 1) organic *causes*; 2) mechanical (or physical) *causes*; and 3) *correlations or covariations*, with no causal connection between or among the variable factors stated or implied.[10] The third group comprises the set of explanations that are better understood as "laws" that do not identify "causes."

Because Darwin did not make these distinctions, some misunderstanding has arisen about his beliefs about the causes of variation.[11] He mixed together causal explanations of both kinds, and both of these with statements about correlations, seemingly as if all three addressed the same kind of question or were of the same epistemic value. But when all three factors are weighed in the same balance it becomes clear that the three factors assume a hierarchical order of explanatory importance in the structure of Darwin's theory. Most important in explanatory value are organic causes; after them, mechanical/physical causes; and last, covariations, see Figure 3.1.

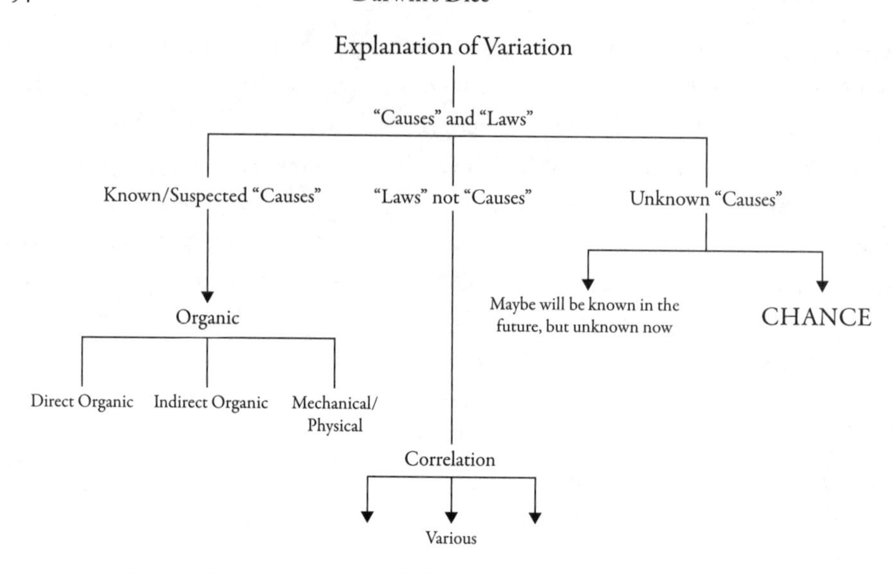

FIGURE 3.1 Perhaps the hardest question of all for Darwin was: how do we "explain" variation? He had some good ideas (the left-hand side of the diagram), some ill-formulated ideas (confusing "laws" with "causes" in the center—these are "correlations" rather than "causes"), and some ideas where he could only say "I do not know" or "just by chance"—not synonymous expressions (right-hand side of diagram).

ORGANIC CAUSES

Organic causes are those causes that work "directly" or "indirectly" on organic structures (hence, organic causes). The only causal factor that works this way identified by Darwin is "physical conditions of existence" (i.e., the environment). Sometimes these conditions induce variations directly on organic structures. If a colder climate "causes" longer hair, or if more sunlight "causes" brighter colors, Darwin claimed the "external" physical conditions were "directly" responsible. This "cause" of variation has come to be known as Geoffroyism, named after the French zoologist Étienne Geoffroy St. Hilaire. Darwin always gave "some" credit to this view, but never much.[12]

Sometimes such "direct action" worked through a different causal pathway, what is usually called Lamarckism: conditions "cause" changes in "habits," and changed habits give rise to new organic structures. The iconic example, though not Lamarck's preferred one, is the long neck of the giraffe. How, we might wonder, did these marvelous necks come about? Earlier theorists had argued that they were an original separate creation. Lamarck argued that the long necks were a product of transmutation ("evolution" was not yet a term of currency for describing organic changes). Earlier ruminants with shorter necks came to find themselves in new physical conditions in which low-lying shrubs and plants had died out, say because of an extended drought. Edible leaves started to become available (in the part of the world where the predecessors of giraffes lived) only on branches higher up off the ground. To keep pace with the ascending leaves "incipient" giraffes began to "stretch" in order to reach the foliage (a change in their "habits"). In consequence of this stretching—Lamarck made the mistake, in Darwin's view, of suggesting the stretching was "deliberate" or "willful"—the necks gradually elongated (a change in structure induced by the change in habits), and the longer necks were then passed down to progeny. Over the course of ages such gradual modification of organization gave rise to the production of entirely new species—in this case the giraffe. The whole Lamarckian process is usually called "use-inheritance," or "inheritance of acquired characters."[13]

Darwin's third "organic" causal factor in variation was his own contribution: physical conditions acting "indirectly" to cause organisms to vary through their action on the "reproductive system" of the parents.[14] This is still an "organic" cause of change because it still has the "physical conditions of existence" working on the organic structure of organisms. But now such external conditions are operating "indirectly." They are modifying not living creatures but rather reproductive organs of parents. Such modifications to the reproductive system in turn give rise to modified offspring—the variations we see in progeny of parents both under domestication and in nature. This particular "cause" was, of all organic causes, for Darwin the most important. Changes to the "physical conditions of existence" give rise to variations through the action of conditions on organs and organisms—not "directly," as cold temperature causing longer fur, but "indirectly," through causing changes in "reproductive organs" or "the reproductive system," these in turn giving rise to variant offspring.

MECHANICAL/PHYSICAL CAUSES

Different from organic causes are "mechanical/physical" causes.[15] These causes are "mechanical/physical" rather than "organic" because they connect a modification in one part of an organism to a modification in another part. They are "causes" rather than mere correlations because one sees a temporal sequence: some part undergoes change and that change precedes another change that evidently results from the first. "Changes of structure at an early age will generally affect parts subsequently developed" is one such example given by Darwin (*Origin* 318: V. 293). Another is, "Modifications in hard parts and in external parts sometimes affect softer and internal parts" (*Origin* 318: V. 291; *Descent* ch. 4, 113). Some parts may become larger as a consequence of drawing nourishment from other parts (*Origin* 318: V. 292).

CORRELATIONS OF GROWTH

Finally, some laws of variation are not causes at all, but only correlations. Darwin created an entire category of cases of variation that he called "correlations of growth," to which he assigned great importance. This category is itself composed of "many laws," some of which we are "utterly unable to understand." To complicate matters, some correlations *are* causes, such as in the examples given in the preceding paragraph, but some are not. For example, rudimentary organs tend to be variable, not "because" changes in other organs cause them to vary but because, being useless to the survival of the organisms to which they belong, they are overlooked by natural selection, and so left free to vary (*Origin* 318: V. 296; *Descent of Man* vol. 1, ch. 2)[16]. In this example no "cause" of variation is assigned, only conditions under which variation occurs. Similarly, "homologous parts tend to vary in the same way, and homologous parts tend to cohere" (*Origin* 318: V. 290). Again, no cause is given, only an observed correlation. In the same group belongs the idea that "specific characters—that is, the characters which have come to differ since the several species of the same genus branched off from a common parent—are more variable than generic characters, or those which have long been inherited" (*Origin* 318–9: V. 297).[17]

Between causes and laws, "causes" is the narrower concept, but also the more potent in explanatory value. If one has laws, one can assert how various phenomena are related to one another, but not *why* they are so related, or which precedes the other in time (an important consideration in causation). A law, for example, might inform us that whenever x is present, y is also present. A cause, by contrast, says not only this much, but also that the *reason* y is present is *because* x is present, *and* that x came first in time *and* that x somehow brought y into existence. Darwin believed that when one has a cause in hand, especially a *vera causa*, one has achieved the highest goal of science—explanation, and possibly even the ability to make predictions.[18]

Causes are sometimes thought of as giving "necessary and sufficient" conditions for an observed phenomenon. "Laws" do not necessarily do so. A canonical example of a necessary and sufficient test is the "cause" of fire. Necessary conditions include fuel, oxygen, and kindling temperature: all must be present for fire to occur. The three conditions are also sufficient: add the three conditions together and one will get fire; nothing else is needed. But a weaker version of this causality thesis exists: a particular effect that requires various prior conditions but without the requirement that these prior conditions are unique to the production of the effect. In the example above, the weaker version of causality would still be a theory of causality if it were true: (1) that several conditions are necessary and sufficient for the production of fire; but (2) that several alternatives for one (or more) of these factors would do the trick as well as others. For example, if fire could be caused in the presence of oxygen, but just as well in the absence of oxygen but in the presence of some other element, then we would still have a causal explanation, but one more open-ended than one in which *only* certain specifiable conditions and *no others* could produce the effect. In such cases, one can preserve the higher standard of "necessary and sufficient" by spelling out all the operative possibilities: "either oxygen, or hydrogen, or x, y, or z" (what Herschel termed "residual phenomena"—see discussion of Herschel in the previous section). If the list of possible operatives were exhaustive, we would still have a "necessary and sufficient" test, even if a less tidy one than when only unique factors with no possible substitutions are in play.

Darwin's views about the causes of variation belong to the "weaker" view of causation described above. Variation, he believed, has no single set of conditions that leads to it. It might be "caused" by a + b, but it might be "caused" by c + d instead. On top of that, variation, he thought, is also "governed" by "laws" that are sometimes not causes, yet laws too, as much as we are able to glean them, need to be included in an account of variation. Making matters even more complicated is the fact that Darwin knew he could not give an exhaustive list of all the causes and laws. "Our ignorance of the laws of variation is profound," he wrote.

NATURAL SELECTION

In the context of causes and laws of variation one may well wonder where natural selection fits in. Is it a cause of variation? Is it a law of variation? If it is not, why does Darwin often call natural selection a "cause of change," even "the paramount cause," surpassing in importance direct and indirect causes (cf. J. Beatty, 2010, 22–44).

As long as the phrase "cause of change" is not qualified so as to align it with "cause of variation," Darwin's language is unobjectionable. Natural selection does, of course, play a role in the modification of species. It can even properly be said to play the "paramount" role. Once variations are given "by nature," natural selection "decides" which variations will be preserved in the struggle for life and which shall perish. The "power" of this natural force comes out most clearly in the 1842 "Sketch" and 1844 "Essay." I quote from the latter:

> Let us now suppose a Being with penetration sufficient to perceive differences in the outer and innermost organization [of organisms] quite imperceptible

> to man, and with forethought extending over future centuries to watch with unerring care and select for any object the offspring of an organism produced under the foregoing circumstances [i.e., variations, induced mainly by physical conditions acting on the reproductive system of its parents]; I can see no conceivable reason why he could not form a new race (or several…) adapted to new ends. As we assume his discrimination, and his forethought, and his steadiness of object, to be incomparably greater than those qualities in man, so we may suppose the beauty and complications of the adaptations of the new races and their differences from the original stock to be greater than in the domesticated races produced by man's agency.… With time enough, such a Being might aim at almost any result.… Seeing what blind capricious man has actually effected by selection during the last few years, and what in a ruder state he has probably effected without any systematic plan during the last few thousand years, he will be a bold person who will positively put limits to what the supposed Being could effect during whole geological periods (*Evolution by Natural Selection,* 115–6; parallel passages in the 1842 "Sketch," 45; and *Origin,* 163–4: III. 7–13).

Darwin specifically states in these passages that by "a Being with penetration sufficient" to cause species to change he does not mean "a Creator." In the "Sketch" and "Essay" he does allow the possibility that Natural Selection, as a "law of nature," may have been fashioned by "a Creator," one who chose to operate through "secondary means," or "laws" (here he uses the word "laws" as "forces," not "propositions"). But Natural Selection, even though cast as a "Being," is not really a Being at all. It is a "law" *and* a "cause," and as such is blind and mechanical. What Darwin wanted to emphasize in the quoted passage (and elsewhere) is the "power" of Natural Selection. Give it the raw materials to work upon, and it can fashion results "without limit."

Natural selection is not, however, a "cause" or even a "law" of *variation.* Yet Darwin often seems to suggest it is, opening the way to potential confusion (see note 11 in this chapter). He does so by placing "natural selection" side by side with other factors in evolution that *are* causes of variation, as if to say that, among several rival causes of variation, natural selection is one of them, even the most potent one. For example, at the end of Chapter I of *Origin*, after recapitulating the major "causes of variability," as he understood them, he concludes by saying:

> Over all of these causes of Change, I am convinced that the accumulative action of Selection…is by far the predominant Power (*Origin* 119: I. 311–22; see also *Origin* 730: XIV. 71–6).

Or again, when contrasting the effects of the acknowledged "causes" of "use/disuse" in producing variation, Darwin sums up by saying:

> We should keep in mind... that the inherited effects of the increased use of parts, and perhaps of their disuse, will be strengthened by natural selection. [How] much to attribute in each particular case to the effects of use, and how much to natural selection, it seems impossible to decide (*Origin* 253: IV. *VII. 382.65.0.165–7*; similar expressions at 282: V. 49; and 284–5: V. 63–7).

The impression is that Darwin regarded Natural Selection as another "cause of variation." This is not so. But it is one cause, among others (including causes of variation), of *change*. It enters the process of species modification *after* variations have been given, through whatever causes.

Darwin came to see the potential confusion in later editions of the *Origin*.[19] In the third edition he added the following remark:

> Several writers have misapprehended or objected to the term Natural Selection. Some have even imagined that natural selection induces variability, whereas it implies only the preservation of such variations as occur and are beneficial to the being under its conditions of life" (*Origin* 164–5: IV. 14.1–2*c*).

He continued to make revisions to subsequent editions. Many passages addressing variation and natural selection were simply deleted. Some others were modified. An important change was made to the fourth edition. Here Darwin made it clear that the "two elements of change" are "essentially distinct." The first of them is "physical conditions causing the variability" and the second is "selection accumulating the variations in certain definite directions" (*Origin* 279: V. 33. 2. d). Even greater clarity comes in the fifth edition:

> We clearly see that the two elements of change are distinct; the [physical] conditions cause the variability; the [selecting agency] accumulates the variations in certain directions, and this answers to the survival of the fittest under nature (*Origin* 279: V. 33.2.e).

Darwin had not misspoken in earlier editions, he had only spoken somewhat carelessly. He commingled "natural selection" with true "causes" of variation in his discussion of "causes of variation" (e.g., *Origin* 318: V. 294, 296). But when one reads closely one sees that Darwin did not ever claim that natural selection caused variation, only that it contributed causally to "organic change." This contribution, we now see more clearly than before, was that natural selection could "accumulate" variations given through other causes.

GEOFFROYISM AND LAMARCKISM IN DARWIN'S THOUGHT

Of the three types of causes described above that are important in explaining variation Darwin attached the greatest importance to what I have called "organic causes." He identified only three of these: "the direct effects" of external conditions (biotic and abiotic) on habits and organization; the "indirect effects" of changed conditions on the reproductive systems of organisms; and what he often called "chance," by which he usually claimed to mean "cause unknown" (recall that "correlations" are not "causes"). It is to be observed that whenever "physical conditions" act "directly" or "indirectly" on organisms, two factors are always involved: the nature of the conditions and the nature of the organisms. Darwin became convinced, but not until the mid-1860s, that of these two factors, the "nature of the organism" is of much greater importance than "the nature of conditions" (*Origin* 78–9: I. 9.1–3e; 81–2: I. 28e). This discovery reinforced his belief that "indirect" causes are more important than "direct" causes because of the greater susceptibility to environmentally induced changes of reproductive organs than of features of creatures already living (*Origin* 80: I. 13).

"Direct effects" has two variants. The first might be called "Geoffroyism" and the second "Lamarckism." The former takes its name from the great nineteenth-century French zoologist Étienne Geoffroy St. Hilaire (not to be confused with his son Isidore). The latter is the theory of evolution propounded by another famous French zoologist, J. B. Lamarck. Both men published their more important "transmutationist" works in the early decades of the nineteenth century, and Darwin was familiar with both by the time he began his *Notebooks*.

The difference between these two theories is subtle but significant. According to Geoffroy, the physical characteristics of species could be influenced "directly" by the action of environmental conditions on organization. For example, a colder climate might induce organisms exposed to it to grow longer hair; the shells of some organisms exposed to more sunlight might, in direct consequence, acquire brighter colors; organisms having access to greater quantities of food might thereby be modified in decided ways, such as becoming larger. Geoffroy presumably believed that such acquired modifications might be heritable, leading to the gradual transformation of organisms.[20] He did not think, however, that such modifications could cause species to mutate into new species, only that they could give rise to variations, or new races.

Lamarck's theory, more famous as a precursor to Darwin than Geoffroy's (thanks, no doubt in large measure, to Darwin's frequent allusions to Lamarck's work as compared to his relative silence about Geoffroy's), was that changed conditions could give rise to, or "cause," modifications in structures "directly," but in a different way than Geoffroy believed (I am here using Darwin's language). The way this cause works, to abbreviate, is that changed environmental conditions—say a new climate or a new food source or a new organic competitor—would cause a change in an organism's "habits," or as we would say today, behavior. The new behavior would induce gradual changes in organic structures, and these structural changes would then be passed down to offspring. We illustrated this mechanism earlier with the example of the giraffe.

Darwin was never much of a Geoffroyian.[21] While he could believe that physical conditions *might* act directly on organization he always insisted that this factor was of slight importance in causing organic change, at best:

> How much the direct effect of difference of climate, food, &c., produces on any being is extremely doubtful. My impression is, that the effect is extremely small in the case of animals, but perhaps rather more in that of plants (*Origin* 276: V. 13–14).

That was Darwin's position in the first edition of the *Origin*, and while he did tinker with the mode of expression in later editions (for example, in the third edition he removed the modifier "extremely" before the word "small" in the second quoted sentence, suggesting possibly a slight shift toward Geoffroyism), he was consistent throughout life in giving this cause the smallest role among the "organic" causes of variation.

More important to Darwin, and to his readers, was the Lamarckian factor of "use-inheritance." The question of Darwin's intellectual debt to Lamarck is a tangled controversy, most of which will be treated later.[22] But as a preliminary, two questions must be addressed. The first is, why did Darwin give a greater role to the Lamarckian factor of "use-inheritance" than to the Geoffroyian factor of what we might call "variational inheritance"? The second is whether Darwin became "more Lamarckian," as many scholars believe, as he continued to ponder the causes of variation and inheritance over the course of many years of reflection.

As to the first, the probable answer is that Geoffroy was not an "evolutionist," whereas Lamarck was. Neither Geoffroy nor his followers believed that existing species could become new species through direct environmental influences. What such influences could produce, they all argued, was new variations, but not new species. This conclusion brought Darwin up short. He had been convinced from his earliest thoughts on variation that variations are incipient species, and anyone who would not allow that incipient species could become new species in time was not worth a lot of attention.

By contrast, Lamarck was a species evolutionist. While Darwin eventually came to reject the Lamarckian explanation for the giraffe's long neck, he retained a role for use-inheritance throughout his life. The reason is traceable to Darwin's observations and experiments. *His* iconic case was not the giraffe but the duck. On the hunch that Lamarck might have been correct about the role of "habit" in causing variation, Darwin experimented on domesticated ducks. This was a preoccupation of his over many years. What he claimed to find was that domesticated ducks tended to have smaller wing bones and larger leg bones than their undomesticated cousins. This, he believed, was proven through measurement of the weights of the respective bones.[23]

What could account for this? If Lamarck was right, an explanation was available in the argument from use-inheritance. Domesticated ducks (Darwin experimented on several varieties) do not fly as much, and walk more, than the same species in nature. The "habit" of not flying, in other words the "disuse" of the wings (and corresponding "use" of legs), could explain the gradual diminution in size of wings and increase in size of legs as compared with birds in nature. If Darwin's empirical observations were correct—and he was sure they were—Lamarck's theory gave the best explanation. He retained the example through every edition of the *Origin* and through both editions of *Variation*.[24]

But we should not exaggerate Lamarck's influence on Darwin. Use-inheritance, though more important than Geoffroyism, always remained a subordinate cause, less important than indirect effects of conditions on the reproductive system. (The question whether Darwin came to assign more weight to use-inheritance later in life will be taken up in chapter 8 below.)

Besides, Lamarck's theory had problems of its own that caused Darwin to downplay Lamarck as a precursor. For one thing, Lamarck believed in organic "progress." When a new species comes along it replaces prior species because it is an "advance" in the scale of nature. Darwin did not accept that idea in the *Origin*, and in fact had rejected it earlier on, in the *Notebooks*, as being not in accord with the "facts." The facts show that creatures often go backward or sideward in their evolutionary changes (e.g., *CDN* E-95).

Secondly, Lamarck's theory required one to believe that many animals are able to modify their habits through "willing," even "conscious willing." This idea too was a non-starter for Darwin. Plants that also evolve have nothing that could be called "will," let alone "conscious will." But even with respect to animals, many are so low in the scale of organization—including Lamarck's special area of study, the lowly invertebrates—that they could not be presumed to have any will or conscious will either. Yet even they, on Lamarck's understanding, evolved.

Finally, much of Lamarck's theory hinged on the idea of new replacements coming into being to replace older lines that evolve into new lines. Older lines do not become extinct, they only change. Darwin did not accept that. Extinction is part of the history of life on this planet. Also, Lamarck argued that the new replacements that filled spaces left vacant by species that changed into new species came about through "spontaneous generation." Occasionally, lifeless matter would spring into life when openings became available in the economy of nature. Darwin's problem with spontaneous generation was that absolutely no evidence could be brought forward in its defense. No one had ever seen lifeless matter come alive, and experimenters who had claimed to produce results to show this could happen were, Darwin thought, careless and unreliable.

DARWIN'S THIRD CAUSE

Of the several causes of variation identified by Darwin this third cause—indirect action of conditions on reproductive organs—was always the most important:

> I have remarked in the first chapter—but a long catalogue of facts which cannot here be given would be necessary to show the truth of the remark—that the reproductive system is eminently susceptible to changes in the conditions of life; and to this system being functionally disturbed in the parents, I chiefly attribute the varying or plastic condition of the offspring (*Origin*, 276: V.8; cf. *Origin* 79: I. 12).

An indirect cause is qualitatively different from a direct cause. One may observe, test, and measure the effects of direct causes. For example, a scientist equipped with a thermometer, an instrument for measuring length of wool, and several species of wool-bearing mammals in cold climates can read on his instruments whether colder climates correlate with longer fur. If they invariably do, one can surmise that the climate "caused" the variation. (It would be consistent with the observed facts to conclude that the longer fur "caused" the colder climate, and to foreclose that possibility, implausible as it may *seem*, one would need to conduct rigorous time-sensitive trials: which came first, the colder climate or the longer fur; and one would need to provide plausible explanations: if the longer fur came first, *why* would it cause colder climates?)

Matters are not so straightforward with "indirect causes" of variation. The reason for this is that indirect causes operate, according to Darwin's account, on the "reproductive system." The basic idea here is that changed conditions act not just directly to change the organization of living creatures, but also indirectly, through some disruption of the normal processes of reproduction. The disruption occurs because of the "extreme sensitivity" of the reproductive organs to environmental conditions. Any change in physical conditions, no matter how slight, is likely to be followed by variations in the offspring of creatures exposed to them.[25]

The introduction of changes induced by changed physical conditions to organisms' reproductive systems opens up an entirely new way of thinking about causes. "External conditions" still constitute a "necessary condition" for variation, as was true for the "direct causes" argued for by Geoffroy and Lamarck. But, unlike the latter, how conditions modify or act on the reproductive system is largely invisible and mysterious. One could not in Darwin's day look inside reproductive organs to examine how they underwent change when subjected to new conditions.

Yet Darwin saw enough evidence, especially in the methodical breeding practices of breeders of plants and animals under domestication, that the exposure of these productions to varying conditions was associated with variations in the offspring. He also had evidence that the association was "causal"; the changed conditions must be acting indirectly on reproductive organs rather than directly on living creatures because the variations would often appear before the environment would have had any time to act directly on already living organisms; or bore no discernable relation to new physical conditions.

Darwin's "third cause" is what, on the score of variation, mainly separates his theory from those of Geoffroy and Lamarck and gives it its distinctly "Darwinian" flavor. To be emphasized here is the point that Darwin's cause of variation operates *before* the act of conception, or perhaps *at* that moment, but not *after* conception has occurred (*Origin* 79: I. 12; cf. 82: I. 26; 276: V. 9).[26] Changes in organisms, even embryos, that result from their *direct* exposure to changing environmental conditions are the sort of changes predicted by Geoffroyian/Lamarckian theories. As we saw, Darwin did not entirely reject those accounts. He kept "use-inheritance" to the end. He even accepted experimental evidence produced by Isidore Geoffroy St. Hilaire that purported to demonstrate that "unnatural treatment of embryo causes monstrosities," and that "monstrosities cannot be separated by any clear line of distinction from mere variations" (*Origin* 79: I. 11). But the third cause, "indirect action," held primacy.

We may wonder why Darwin did not attribute more weight to Geoffroyism and Lamarckism. Or to put it the other way round, why did he prefer his cause, thinking it to be the most important. His reason is straightforward: Geoffroy and Lamarck had theories that could explain *some* variations, but not *all* of them. Living organisms of the same species exposed to very different conditions often vary not at all; and the offspring of living organisms of the same species subjected to nearly identical external conditions sometimes vary a great deal (*Origin* [*Variorum*] 278–9: V. 27–32). Many variations are not helpful, many more are positively harmful to the survival prospects of those acquiring them. If physical conditions work *directly* on organisms to modify them, one would expect the modifications thus induced to correspond to, or reflect in adaptive ways, the changing conditions. But they do not always do so. The *direct* influence of external conditions is thus *not* the whole answer to variation. What one really needs to grasp is that progeny often vary, when they do, *because of* prior changes to the reproductive organs of the parents.[27]

CHANCE AS A FOURTH CAUSE?

What, then, about "chance"? We earlier identified this as yet another cause of variation, in addition to "direct" and "indirect" action of conditions on organization. "Chance" for Darwin was shorthand for "cause unknown." At the same time, however, "indirect action" of conditions on organization was also, in some sense, an "unknown" cause, in the sense that while one may know physical conditions act "indirectly," one cannot say how or why they act as they do. To put it differently, many variations are unpredictable. There is no guessing what might turn up in any given new generation of offspring. My conjecture is that to Darwin the two expressions amounted to the same thing. "Chance" causes are unknown because they work "mysteriously" on the reproductive system; conversely, the "indirect" action of physical conditions on the reproductive system is so mysterious that we may as well say changes induced this way are "by chance." In other words, "chance" and "reproductive system" are systematically related in Darwin's thought. Thus, as he modified his claims about "chance" in subsequent editions of the *Origin* (after the first) and later writings, we would expect him to make parallel modifications regarding his arguments about the "reproductive system." This he does, as can be shown by taking a closer look at "chance" as "cause of variation" in Darwin's thought.

The argument to this point identifies "direct" and "indirect" causes of variation. But it leaves open a question that would prove to be of great interest to Darwin's audience, friends and foes alike: are these variations "guided" or "designed" by a higher intelligence? If they are not guided, then organic evolution could only be understood as a "random" or "accidental" process, and that thought was anathema to just about everyone.[28]

The idea that variations occur through the *direct* action of external conditions on living organisms (the Geoffroyian and Lamarckian ideas) would seem to be more congenial to the idea of a designed creation than the rival idea of *indirect* action of environment on reproductive organs. Lamarck in fact had already argued this, as had Robert Chambers in the *Vestiges*. Apparently it is easier to imagine a divine intelligence creating plants and animals that could adjust to their actual physical circumstances when they are alive and confronting changing conditions than to think such a creative force could adjust the internal reproductive organs to ensure that just the right variants would be produced, and that these adjustments would be made even before conception! The problem in Darwin's theory for the proponents of "design" was much more in Darwin's other cause of variation, the *indirect* action of conditions on the reproductive organs of parents, than in the idea of organisms already alive making necessary adaptations to changing conditions.

This problem was not intractable. It could still be the case that variations indirectly caused are themselves "designed" in reference to the adaptive needs of offspring. On this account, "conditions of existence" could act "indirectly" on organisms to produce variations, while at the same time allow that such variations as do come about through this means are "ordained" to suit the new variants, and ultimately, the new species, to the particular conditions they face in the struggle for life. This is a good approximation of the views of Charles Lyell and Asa Gray, two of Darwin's strongest supporters in the early days after the *Origin* appeared, of variations in nature—that they were foreordained by a higher intelligence to be well suited to a changing environment.[29]

Unfortunately for that understanding, whenever Darwin spoke of *indirect* environmental influences creating favorable adaptations he did not ever suggest that God planned it this way. On the contrary, one of the facts of nature that fascinated him most—and surely helped cast great doubt in his mind about a role for intelligence anywhere in the design process, was the fact that so many harmful designs are to be found in nature. Some creatures are born so ill-adapted that they do not really have any chance at all to survive or at least to propagate. That did not seem to Darwin to reflect intelligence. Darwin was also disturbed by any notion of God that would cast Him in the role of an inept, or worse a cruel, creator. How could a good God plan a world destined to be filled with so much senseless death and evident misery? Darwin was quite sure that was impossible.

In light of these considerations it is easy to see that Darwin faced a dilemma, and he knew it. He could either reject God altogether ("like confessing a murder," he once wrote to Hooker), or accept a God who fashioned laws in the beginning but who was sufficiently unfeeling as not to care whether these laws would produce death and misery; or worse, who must have been intelligent enough to know that this would be precisely the result of such laws. How did Darwin decide?

If one were to examine only one single edition of the *Origin* it would be impossible to draw any certain conclusion.[30] In any given edition Darwin might be found to be arguing for "undesigned laws," or for "designed laws," or for both at once. But the issue becomes more interesting and informative when we look closely at each of the six editions Darwin edited for publication and the changes he made between the first and sixth editions in passages where the question comes up. Darwin took it on indirectly, not by asking whether a higher intelligence designed *all* variations, but whether "chance" was the proper explanation for *any* of them. The two explanations are mutually exclusive, meaning that if one is correct the other must be false.

A designed world in all of its parts and operations cannot be a chance world in any of them; and a world in which chance plays any role at all seems to be one that excludes a place for an omnipotent designer. Eventually, it is necessary to decide one way or another, and Darwin understood this.

But Darwin was always reluctant to state openly that his theory excluded a role for divine intelligence. Such a confession may have brought his theory to ruin among his contemporaries. So he allowed in published works and even in many letters to close friends (except, notably, J. D. Hooker) a possible role for an "indirect" divinity who acted through "intermediate laws." Or else he pleaded that such ultimate questions were beyond his limited capacity to understand, and so he would simply not attempt an answer either way. But most often he simply changed his way of saying things—particularly by getting rid of the word "chance" altogether and simultaneously disguising or obscuring the role of "reproductive organs" in the mysterious process of variation.

NOTES

1. Do species, like individuals, have something like "average life-spans," making them susceptible to extinction even when conditions remain the same? Early in his career Darwin believed this might be the case, but he seems to have abandoned the idea by the time of the *Origin*.

2. Darwin did gain much understanding about geology from his forerunners in this field—especially Hutton, Lyell, and his mentor at Cambridge, Sedgwick. He did not merely read and digest their views, he made his own significant and original contributions. But his eyes were on a bigger prize—organic evolution, a concern that took him from geology to biology. The intricacies of the connections between the two lines of thought have been carefully analyzed by Sandra Herbert, Martin Rudwick, Michael Ghiselin, James Secord, and others. But in the case of geology I think Darwin was building on an already laid foundation, whereas in organic evolution he was forging a fundamentally new path.

3. Ghiselin (1969), 160 disagrees: "It was in no way essential [for Darwin] to explain how, or why, organisms vary. The cause of variation was not relevant to his argument." This is hard to square with the text. Not only does Darwin devote much space to the causes of variation, he changed his mode of presentation extensively over the years, showing that he was concerned to get the exposition correct.

4. *CCD* [February 7, 1857], letter from Darwin to B. P. Brent and editors' footnotes.

5. Ruse (1978) points out that Herschel was not always entirely clear on what he meant by the expression *vera causa*. Literally it means "true cause," but what is a "true cause"? Ruse, in discussing Herschel, associates it with the idea of "proof by analogy" (accepted by Darwin up to the *Origin*), then (for a later Darwin) with "proof by consilience of inductions," an idea Darwin could find in both Herschel and Whewell. Gildenhuys (2004) disputes Ruse's reading here. Thagard (1977) emphasizes "consilience" and Whewell as influences more important than "analogies" and Herschel. I am not sure either of these views is precisely what Herschel meant by "true cause" (although I agree with Ruse that he was not always clear). Here is Herschel in his

own words: "Experience having shown us the manner in which one phenomenon depends on another in a great variety of cases, we find ourselves provided, as science extends, with a continually increasing stock of such antecedent phenomena, or causes (meaning at present merely proximate causes), competent, under different modifications, to the production of a great multitude of effects, besides those which originally led to a knowledge of them. To such causes Newton has applied the term verae causae; that is, causes recognized as having a real existence in nature, and not being mere hypotheses or figments of the mind" (*Preliminary Discourse* 144). This could mean "proof by analogy," but it could also mean "real, observable, measurable forces" as opposed to Hume's "fancies of the imagination" or Kant's "fundamental ideas" (i.e., innate concepts upon which meaningful sense-experiences depend).

6. Here I depart from Gildenhuys (2004), who believes Herschel was distinguishing between *verae causae* (i.e., observable causes), and "mere hypotheses" that are only "figments of the mind," (i.e., that "causes" and "hypotheses" are two different steps done a Herschelian view of scientific method) (2004, 595). The passage from Herschel that he cites (see previous note) seems to me to be making a distinction between "causes that have a real existence" and "causes" that are merely postulated as hypotheticals. But the passage is admittedly difficult.

7. Some philosophers disagree; cf. J. Austin (1962).

8. Darwin was not naïve about causation. He had read David Hume and Immanuel Kant on the nature of "causes," but he showed little interest in their metaphysical quarrel about what a cause "really is," much less in later epistemological issues surrounding the trustworthiness of observations or the warrantability of beliefs based on them. In this respect he was a scientist of his day. The warrant for belief is observation (provided it is careful and verifiable by other observers); and the aim of scientific observation is to discover "true causes," which to Darwin meant necessary and sufficient conditions to explain phenomena. For discussion of Darwin's methodology throughout his career, cf. M. Ghiselin (1969); D. L. Hull (1973a); Ruse (1975, 1978).

9. The distinction between causes and laws in contemporary epistemology and philosophy of science is more complicated than this. I am more interested here in how questions of causality and correlation were understood in Darwin's day, and especially by Darwin himself. Unfortunately, Darwin sometimes used "laws" and "causes" interchangeably, and his analysis of variation suggests that he was not fully aware that the two are different and point to different classes of "facts" in nature. Darwin's primary source for the concepts of "laws" and "causes" was J. Herschel 1831 *Preliminary Discourse*, read and annotated by Darwin in 1839 (see *CDN* N-60). W. Whewell (1840) was another important influence.

10. Darwin drew another line of demarcation among causes" between "indefinite" and "definite," but only in late editions of the *Origin*. The distinction appears to center on the question of whether variations are "predictable in principle" ("definite") or "unpredictable in principle" ("indefinite"). I will not engage this distinction here because, as I see it, both classes are "chance" variations for Darwin. Discussions in Lennox (2004) and Beatty (2006).

11. A recent example is Gildenhuys (2004, 597–8), who believes Darwin regarded as "causes of variation" all of the following: correlations of growth, heredity, reversion, natural selection, and perhaps even more. He does not notice the distinction developed here between laws (all of the above) and causes (only three, including *none* of the above—see below). But it is hard to find fault. Darwin did not see the distinction clearly himself!

12. Darwin continued through late in life to give the smallest role in variation to Geoffroyism (e.g., *Descent of Man* [1872], ch. 4). Cf. R. J. Richards (1989, 37–8), who argues that Buffon and earlier writers were responsible for introducing the Geoffroyian idea into scientific thought. Darwin was apparently unfamiliar with these earlier writers. Further discussion of Geoffroy's apparent influence on the early Darwin in Richards (ibid., 86–9). Cf. Johnson (2007) for Darwin's shifting views about Buffon and admitted ignorance about earlier writers.

13. "Use/disuse" and inheritance of acquired characters ("Lamarckism") was retained in the relatively late *Descent of Man* (1872, vol. 1, ch. 4), but with no noticeable change in its relative importance in accounting for how variations arise or in how "giraffes" are implicated. Buffon had argued that giraffes were a class unto themselves with no known conspecifics (along with nine other species of animals and four other orders composed of multiple families, genera, and species). Cuvier believed giraffes were created in the beginning as a sui generis species, fitted for a particular habitat, related to other Ruminata but certainly not descended from other species.

14. This "cause" shows a resemblance to David Kohn's view (1980, 67–170) that Darwin believed "sexual reproduction" was, in the *Notebooks* at least, the main mechanism of species alteration (in addition to the operation of the environment on organization and Lamarckian use-inheritance). R. J. Richards (1989, 124–6), takes issue with Kohn on the grounds that he finds Darwin to reject the idea that modifications come about in "*full-grown* individuals." I think Richards may be misreading Kohn here because Kohn does not tie new adaptations to alterations of adult individuals but rather to "the habit of sexually reproducing organisms to produce peculiarities, i.e., to vary" (Kohn, 132). Both Richards and Kohn overlook the close connection in Darwin's thought between "indirect action of conditions on reproductive system" and "chance variation," as developed below.

15. Darwin did not employ the *phrase* "mechanical causes" in the *Origin*, but he did use it in the *Descent of Man* (1872, vol. 1, ch. 4, 113), in exactly the sense intended here: "[Causes of variation include] the effects of the mechanical pressure of one part on another; as of the pelvis on the cranium of the infant in the womb." The example remained the same from the *Origin*, so the idea is not new with him in 1872, only the expression.

16. "Abortive organs" had earlier been identified as "caused" by what Darwin called "absorption" in *Notebook* D: "There is probably a law of nature that any organ which is not used is absorbed.—This law acting against hereditary tendency causes abortive organs" (*CDN* D-166). By the time of the *Origin* Darwin had dropped the idea that a causal factor in abortive organs was in play.

17. Perhaps Darwin believed his "laws" were all "causal," given the proper explication. For example, one type of covariation is when a hard structure in early development "causes" another, softer structure to undergo modification during growth from its earlier form. One can then say that the "hard part caused the shape of the soft part." But that sort of translation does not work so well for other "laws." Darwin would not want to say, I suspect, that the white fur of the cat "caused" its blue eyes or its deafness (*Origin* [*Variorum*] 84: I. 42). The various features just always "go together." Most of Darwin's "laws" are of the latter type. Beyond that, when he did refer to "causes" under the name "causes," he always referred explicitly to external physical conditions "causing" organic change; or to changes in one part "causing" changes to another part.

18. "Although no new fact be elicited by these speculation[s] even if partly true they are of the greatest service, toward the end of science, namely prediction.—Till facts are grouped & called

there can be no prediction.—The only advantage of discovering laws is to foretell what will happen & to see bearing of scattered facts" (Darwin, 1838: *CDN* D-67).

19. Some of Darwin's most careful and sympathetic readers were confused about the distinction between variation and natural selection, including J. D. Hooker (*CCD* vol. 7, 437 [December 20, 1859]; vol. 8, 238 [June 5, 1860]); Charles Lyell (*CCD* vol. 8, 400 [September 30, 1860]; vol. 8, 403 [October 3, 1860]; vol. 13, 22 [January 16, 1865]); Hugh Falconer (*CCD* vol. 11, 15 n. 6; cf. 7 n. 10; vol. 11, 36 [13] January 1863]); and W. H. Harvey (*CCD* vol. 8, 371 [September 20–24, 1860]). Darwin later came to admit that he was not always as clear in expression as he should have been (e.g., *CCD* vol. 8, 496 [November 26, 1860], letter to Asa Gray). Darwin's engagement with the issue—and with his correspondents—comes out most clearly in his reiterated attempts to develop the "stone house" metaphor, discussed in a later chapter.

20. Darwin included a reference to the work of Geoffroy in his "Historical Sketch," appended to the first American and second English editions of the *Origin,* and retained thereafter, with further additions in subsequent editions (for details see Johnson, 2007). In it he credited Geoffroy with having "suspected" that "conditions of life" cause "various degenerations" in species, without, however, suggesting that species actually change into new species (a presumption affirmed by Geoffroy's son Isidore in his "Life" of his father, also cited in the "Historical Sketch"). In his reference to Étienne Geoffroy Darwin also drew attention to the remarkable coincidence that three men—Goethe, Erasmus Darwin, and Geoffroy—had actually partially anticipated his own theory in precisely the same year, 1794–1795 (*Origin* 60–1: "Historical Sketch" 13–7).

21. In all editions of the *Origin* and all the way through the *Descent of Man* Darwin's position was consistent: "direct action of conditions on structures" may play *some* role in variation, but never much, and certainly much less than other factors (cf. *Descent*, vol. 1, chapter 3). It is not clear from his writings that Darwin even recognized Geoffroy as an influence, and it was not until more than 100 years later that this strand in Darwin's thinking was differentiated from the more noticeable "Lamarckian" strand, by Ernst Mayr in 1963 (Mayr, 1963, [1988]). The authorities Darwin cited for the Geoffroyian view included E. Forbes, Gould, Wollaston, and Moquin-Tandon (*Origin* 277: V. 16–19). He did not cite Geoffroy himself. Like Geoffroy, these other authorities were not transmutationists. All of them believed that new races might be produced through the direct action of external conditions, but not that new species might thus be created. Some modern authorities believe Lamarck was actually the source of both Darwin's Geoffroyism and his Lamarckism (e.g., Kohn and Glick, 85).

22. These will be discussed in chapter 8 below. Lamarck is the first author to whom Darwin devoted any real attention in the "Historical Sketch" (see n. 9 above). He credited Lamarck with being the first to "arouse attention to the probability of all change in the organic world," drawing notice specifically to Lamarck's discussion of the "direct action of physical conditions" on organization (the Geoffroyian mechanism), and to "use, disuse, and habit." But immediately thereafter Darwin criticized as incorrect Lamarck's "law of progression," saying in a footnote that his opinions were founded on "erroneous grounds" (*Origin*, 60, "Historical Sketch" 6–13).

23. Darwin had a lifelong interest in the question of what happens to ducks when they are domesticated: do their leg-bones grow larger and their wing-bones become smaller? He resolved to experiment on ducks as early as 1839 (*CDN* 493 ["Questions and Experiments"]), and by the time of the 1842 "Sketch" he had decided tentatively that "when an organ is not used it tends to diminish (duck's wing?)" (*Evolution By Natural Selection*, 82; cf. "Essay," 238). The example

remained a fixture in his published writings from that point on: cf. *Origin* 83: I. 34; *Variation* vol. I, 299–302, in which tabulated results of his experiments are given.

24. Ducks were not Darwin's only example of Lamarckism "use-inheritance." Other favorites were the drooping ears of domesticated pets like dogs (disuse), the enlarged intestines of domesticated pigs (use), and even perhaps the enlarged arms of blacksmiths (use).

25. Darwin admittedly did not have a great deal of observational evidence for variations being caused by environmental influences on reproductive organs, although he did try to give some (e.g., Peckham, 1959, 181–2 [6th ed.], where he discusses carnivorous birds and exotic plants bred under domestication). His argument at this point is mainly hypothetical and inferential: if observed changes *cannot* be attributed to *direct* environmental effects, then it is plausible to think they are attributable to *indirect* environmental effects. For a strict empiricist (as I think Darwin was) it would be sufficient that the stipulated "cause" would be "observable in principle." (I ignore here Hume's contention that causes per se are non-observable even in principle; one only observes unvarying sequences and infers causal relationships.)

26. M. Ghiselin (1969, 164–74) draws attention to the important role of what he calls "developmental mechanics" in Darwin's views on variation: that is, to changes that occur anywhere from pre-conception to mature adulthood as part of regular developmental processes (personal correspondence). By "reproductive system," he argues, Darwin meant "the totality of mechanisms producing new individuals," including mechanisms that operate after conception, "not the more restricted usage to which we are accustomed." But in the first edition (deleted in later editions), Darwin is explicit: variation is usually caused "before the act of conception."

27. Darwin did not hold an entirely consistent position about the importance of "direct action" of physical conditions in causing variation, even across the six editions of the *Origin*. In the first edition he stated that "we cannot tell," but that certain "considerations" inclined him to "lay very little weight on the direct action of the conditions of life." By the fifth edition, instead of laying "very little weight" on direct action he was weighing "not much weight," and by the sixth edition he was only laying "less weight" on this factor than on other "causes of which we are quite ignorant." Despite such changes as these, however, "indirect action *always* remained a more important cause" of variation than "direct action."

28. Gavin de Beer, "Forward," in *Evolution by Natural Selection*, 16. To the evidence previously cited for this assertion I should add Herschel 1831 and Whewell 1840 (discussion in Ruse, 1978). Whewell choked on chance more than Herschel, so much so that he forbade any copy of *Origin* to be catalogued into the library of Trinity College. As Ruse says, Herschel (unlike Whewell) did his best to accept Darwin's theory, always swimming upstream (unsuccessfully) against its implications of "no design."

29. Charles Lyell (1837, vol. 2, 388–92): "We must suppose that when the Author of Nature creates an animal or plant, all the possible circumstances in which its descendents are destined to live are foreseen, and that an organization is conferred upon it which will enable the species to survive under all the varying circumstances to which it must inevitably be exposed" (reproduced in *CDN* 256, C-53-1; cf. *CDN* 426, n. 105-3). Darwin recorded his doubts about this view at Notebook C-53: "There must be some sophism in Lyell's statement."

30. The two editions most commonly found in libraries and bookstores are the first (1859) and the sixth (1872). But Darwin made changes, often substantial ones, to every edition after the first. It is worth consulting all of them to gauge the evolution of his thinking about his theory over the 14 years that he continued to revise. A convenient way to do this is to consult the *Variorum* edition compiled by Morse Peckham in 1959.

4

Chance, Nature, and Intelligence

NATURE

What drove Darwin to chance as a source (cause) of variation? Alternatives were certainly available, even ones that did not involve an intervening divine agency. In particular, Darwin could have fallen back on the idea that "nature" causes variations. It was not an uncommon idea. Ever since Aristotle and right up to Darwin's day, "Nature" could be, and often was, cast as an intelligent designer. Darwin, perhaps without thinking it through initially, adopted similar language and expressions. But did he ever really believe it?

Many early readers of the *Origin* thought that he did. Admittedly, Darwin was fond of expressions that cast "nature" as a conscious agent: "nature gives successive variations; man adds them up in certain directions" (*Variorum*, 1959, 105); "[Man] can never act by selection, excepting on variations which are first given to him...by nature (112); "Useful variations [are] given to [man] by the Hand of Nature" (145; see also 167). In later chapters we find nature "providing against" structures that are ill-adapted for survival, for example, by "giving to trees" a tendency to bear flowers with separated sexes (189), "grant[ing] long periods of time" for the work of variation and natural selection (192) "making beautiful productions" (370), "work[ing] by a method" (403), even perhaps "condescending to the tricks of the stage" in causing perplexing variations to occur (667) or "taking pains to reveal her scheme of modification" (747). Summing up his argument in Chapter 14, Darwin wrote: "Variability is not actually caused by man; he only unintentionally exposes organic beings to new conditions of life, and then nature acts on the organization and causes it to vary" (30).

Any power that "gives," "provides against," "makes beautiful productions," "grants" (as in gives permission), "works by a method," "condescends," "takes pains," and above all "acts" sounds very much like an intelligent power. It is easy to forget, in the face of so much "metaphorical" usage, that Darwin intended his usage to be understood as only metaphorical throughout. In the larger tradition of biological thought that Darwin inevitably inherited from earlier natural philosophers, nature *was* understood as an intelligent agent. This view of nature goes back to the Greek philosophers, particularly Aristotle.[1] Aristotle conveyed this understanding in a number of nearly immortal phrases: "Nature does (or makes) nothing in vain," "Nature acts for the best, either always or for the most part," "Nature makes nothing that is superfluous," and so forth.

If that is a view of nature that Darwin wished to reject, why did he so frequently employ terms whose likely result would be to perpetuate a grand misunderstanding? Many of Darwin's readers took his meaning in the Aristotelian sense. Indeed, Darwin was criticized after the *Origin* appeared for using expressions like these, as they seemed to point to an intelligent "maker." This was the point Wallace made to Darwin in 1862, when he pointed out that the term "natural selection" seemed to "personify" nature as an intelligent agent.[2] The phrase was "too anthropomorphic," as Ruse put it (Wallace, quoted in Ruse, 2003, 123). Wallace was the one who urged Darwin to replace the term "natural selection" by "survival of the fittest," to rid it of its teleological connotations. It was a substitution Darwin made in the fifth edition of the *Origin* (*Variorum*, 1959, 145, 164, 168, 176, 179, 202, 223, 228, 271, 279, 344, 362, 561, 733).

But Darwin understood all along that Nature is not a conscious or intelligent agent.[3] He made the point explicitly in the third (and subsequent) editions of the *Origin* with regard to "natural selection." The expression was always intended metaphorically:

> It has been said that I speak of natural selection as an active power or Deity; but who objects to an author speaking of the attraction of gravity as ruling the movements of the planets? Everyone knows what is meant and is implied by such metaphorical expressions; and they are almost necessary for brevity. So again it is difficult to avoid personifying the word Nature; but I mean by nature only the aggregate action and product of many natural laws, and by laws the sequence of events as ascertained by us (*Origin, [Variorum]*, 165: IV. 14. 6–9: *c*).

Darwin rejected any account of transmutation that made reference to a "spark of intelligence," a "vital force," an "interior sentiment," anything, in short, that was beyond observational scrutiny or that implied consciousness. Such answers are, he claimed, not explanations at all, but merely rephrasings of the original object to be explained.

NATURAL LAWS

A better answer, and one that had greatest appeal to Darwin, is that evolution proceeds according to "natural laws," some of which, he confessed, were not well understood but could be assumed to be operating throughout biological nature at all times, just as the law of gravity, not understood until Newton, operates throughout physical nature. He thus rejected almost from the beginning a role for a directly intervening creative force. The search for such a force was both futile and misdirected. The search of the scientist should be for laws:

> **The Grand Question, which every naturalist ought to have before him, when dissecting a whale, or classifying a mite, a fungus, or an infusorian, is "What are the laws of life"** (*CDN* B-229, emphasis in the original).

The search for laws would "direct our examination" of changes in species and "guide our speculations with respect to past and future" (*CDN* B-228). Speculations with regard to the future, i.e., the ability to make accurate predictions, would assume special prominence in Darwin's thought. The ability to make predictions became almost a hallmark of genuine scientific inquiry. And while Darwin was in many cases never able to predict accurately, particularly regarding the direction of organic change, he nevertheless made the search for laws his guiding principle of scientific activity (cf. *CDN* B-236; E-3; and often). And, we should add, he often *did* predict correctly, even absent empirical confirmation, as in the notorious case of his belief that the 12-inch nectary in the orchid *Angraecum sesquipedale* (the Madagascar star orchid) would probably require by his theory an insect with a proboscis of at least that length for cross-fertilization by insects to occur. Such an insect had never been seen to Darwin's knowledge, but Rothschild and Jordan found just the candidate in 1903, and, pleasingly, just where it should be found—in Madagascar![4]

But to attribute life's complexities to the operation of natural laws does not necessarily rule out a role for divine creation. One need merely suppose that an omniscient God fashioned "nature" in the beginning in such a way that the divine plan would unfold according to the logic of the laws He had created. In other words, one can save intelligence by simply moving it back one step—from daily, active intervention to a more original, and prescient act of law-making. After the original intelligent creative act, the laws would ensure nature's unfolding in strict accordance with those laws. Such seems to have been the view of Newton, for one, an important influence on Darwin. For Newton, heavenly bodies course through their heavenly paths as they do, not because the creator presides at every instant of time over the paths they should follow, but because they are governed by physical laws that ensure a proper behavior for each heavenly body.[5] Ruse applies the insight to Darwin:

> Darwin solved [Paley's problem of evident design in nature] by taking the directly intervening God of miracles out of the scene. God was no longer invoked as a direct supplement to science. God may still be designing in the background, but he is now doing it at a distance, through the agency of law (Ruse, 2002, 112).

The idea that nature is regulated by constant laws does away with the need for an actively intervening deity in any of nature's operations, whether regulating the movement of planets or determining the creation of species. But it does not answer a prior question: whence the laws? One might believe that natural laws "just are," and leave it at that. Or, one might try to account for the origin of the laws. Darwin began his reflections on this question by adopting the "deistic" position that the laws were themselves designed "in the beginning" by a creative force:

> The question if creative power acted at Galapagos it so acted that bi[r]ds with plumage <<&>> tone of voice partly American North and South.—**(& geographical <distr> division are arbitrary & not permanent. This might**

> **be made very strong if we believe the Creator creates by any laws, which I think is shown by the very facts of the Zoological character of these islands.**) So permanent a breath [a directly intervening creative force?] cannot reside in space before island existed (*CDN* B-98, emphasis in the original).

The idea seems to be that the only plausible way to account for the presence of non-native species on recently formed islands (like the Galapagos) is that they have been transported there by natural laws rather than "breathed" into life by a creative force after the islands were formed.[6]

But this does not quite settle the issue about where Darwin stood on the central question of intelligence in nature. As Ruse says, "God *may* still be designing in the background." But then again He may not. Ruse, and many others, think that Darwin subscribed to the view that laws are intelligently created, thus allowing him to preserve a fundamental role for an omniscient God in his theory:

> Darwin made it very clear that, far from intending his arguments as a refutation of God's existence, he thought his position confirmed and supported it.... In Darwin's case, we would expect God to create through law, and a cost and consequence of this would be pain and destruction—all for the greater good (Ruse, 2002, 116).[7]

Ruse is right to this extent, that if laws really are at work in nature, it was not a big stretch for many of Darwin's contemporaries to attribute the origin of those laws to an omniscient God. Such a view would seemingly eliminate any role for chance in nature's organic development. Chance would be, as Darwin himself stated, merely a stand-in for "our ignorance." What we attribute to "chance" would actually be the action of unknown "laws," but these laws might themselves have been designed to produce "guided," even "foreseen" results. Plan, purpose, and design could be comfortably accommodated within a worldview that attributed natural productions to the operation of natural laws. The question is, did Darwin accept that view in his own thinking?

The *Notebooks* do not settle the question. At first it seems that he had doubts. Could a creative force, even an "omniscient" one, really have foreseen, for example, all the travel needs of species descended from earlier progenitors? The following entry appears early in the *Notebooks*:

> How does it come wandering birds such as sandpipers, not new at Galapagos? Did the creative force know that <<these>> species could arrive? Did it only create those kinds not so likely to wander? Did it create two species closely

> allied to Mus[cicapi] Coronata, but not coronata? We know that domestic animals vary in countries, without any assignable reason (*CDN* B-100).

But some pages later in the same *Notebook* B, Darwin seems to have reverted to the idea that an omniscient god acted at the beginning by creating laws and then let "nature" take over from there. At least this is what he takes Étienne Geoffroy St. Hilaire to have argued in his 1830 work *Principes de philosophie zoologique*, and he cites Geoffroy with seeming approval:

> E. Geoffroy St. Hilaire says grand idea god giving laws & then leaving all to follow consequences (*CDN* B-114; cf. *CDN* B-114, n. B-114-3 for the passage of Geoffroy's work to which Darwin was referring).

Such "consequences" as implied here include all of the various phenomena associated with the theory of descent with modification—including travel needs of migrating species and the origin of variations—that Darwin believed his theory could account for. To this point they leave the key questions about ultimate causes on the ground of "no directly intervening Creator," but without saying whether a deity might be operating indirectly and remotely on the world through laws that the Creator foresaw would do the necessary work to ensure the unfolding of life on the planet according to a designed plan.

On the question of the existence of a "creative force," then, Darwin was in a muddle. He was certain by 1837 that no "direct intervention" occurred. But he still was uncertain whether such a creative force might be acting indirectly, through "laws." He started to doubt even an indirect role for a Creator later in *Notebook* B, where he observed: "The creative power seems to be checked when islands are near continents: compare Sicily and Galapagos!!" (*CDN* B-160e). The "creative power," if one existed, would be operating in strange ways if flora and fauna on islands somewhat remote from the mainland were similar but not identical to species similar to the mainland species. To suggest that the creative power might be "checked" is a terrible blow to the creative power. Darwin could not have said this if he had been sure that the creative power was all-powerful and all-knowing.

The growing emphasis on the importance of laws in later *Notebooks*, especially D and E, contributes to the impression that Darwin kept moving further away from a belief in divine influence as he continued to ponder. For one thing, if a Creator does govern through laws, it would seem that His power is thereby severely circumscribed. For example, if Erasmus Darwin's theory about sexual organs being the "highest point of organization" of sexually reproducing animals were true, the case of the Mule would be anomalous, and "**The creator would thus contradict**

his own law" (*CDN* D-19, emphasis in the original), and this, presumably, he cannot do.

If the sovereign power of the universe cannot violate His own laws, what kind of sovereign is that? Sovereigns are, by definition, above the law. They may choose to obey the laws they impose, but what is to stop them if they choose not to do so? To allow that laws are sovereign is to deny that God is sovereign, and implicitly to deny that a supreme legislating God exists. Such may well have been the direction of Darwin's thinking, even as early as 1837.

Some passages in the *Notebooks* and the 1842 "Sketch" suggest that Darwin's thinking did not go as far as the logic of sovereignty might suggest. For example, when Darwin first encountered Herschel's famous phrase about the replacement of extinct species by others being "that mystery of mysteries," he clapped his approval:

> Herschel calls the appearance of [all] new species the mystery of mysteries, & has grand passage upon [the] problem [of the origin of new species]! Hurrah.—[He refers to the operation of] "intermediate causes" (*CDN* E-59).[8]

Darwin was sufficiently impressed with Herschel's expression that he included it in the opening paragraph of the *Origin*. Since Herschel was explicitly denying that the Creator could be limited in His activity by his own laws, one might suppose that Darwin's enthusiasm for Herschel's opinion extended even to this aspect of it. But Herschel actually downplays the likelihood that the Creator would ever violate His own laws, and emphasizes instead that "we are led, by all analogy, to suppose that [the Creator] operates through a series of intermediate causes, and that in consequence the origination of fresh species, would be found to be a natural, in contradistinction to a miraculous process" (in *CDN* E-59, n. 2). The "miraculous process" here refers to a directly intervenient God. Darwin agreed that no direct divine intervention occurred. A "natural process" is one governed by natural laws, which are, on Herschel's view, laws made by the Creator. Darwin had doubts that the natural laws were divinely created.

That impression is further strengthened by something else Darwin confided to himself in *Notebook* C. The date of the passage is relatively early, Spring 1838, certainly well before Darwin had read Malthus, but some time after he had discovered he had his own "theory." The passage is interesting for giving a glimpse of what seems to have been an important revelation for Darwin, one that bears directly on the role of a creative force in nature.

> Once grant that <<species>> one genus may pass into each other.—Grant that one instinct to be acquired (if the medullary point in ovum has such organization as to force in one man the development of a brain capable of producing more glowing

> imagining or more profound reasoning than other—if this be granted!!) & whole fabric totters and falls. Study gradation, study unity of type, study geographical distribution, study relation of fossil with recent, the fabric falls! (*CDN* C-76–7).

There can be little doubt that the "fabric" to which Darwin here refers is the whole machinery of divine creation and a purposeful nature. Even humans (in the quoted passage) are not at all special. Their time too will come to an end, "or how dreadfully we are deceived." One wonders where sentiments like these permit any room for an omnipotent and designing deity at all.

Some of Darwin's contemporaries wondered the same thing, especially about where Darwin stood, but were of two minds. Some, like Harriet Martineau, a perceptive contemporary author and a close friend of Darwin's brother Erasmus, detected some verbal cheating in a "theological" direction in the *Origin*: "I think it is a pity that 2 or 3 expressions [suggesting a role for a Creator were] used: but the theory does not require the notion of a creation." But, she quickly added, "my conviction is that Charles D. does not hold it" (quoted in Desmond and Moore, 486).[9] Others, including two of Darwin's earliest supporters within the scientific community, Asa Gray and Charles Lyell, apparently believed that Darwin remained within the bounds of Christian respectability. Gray, for instance, saw a way to have it both ways: variation guided by intelligence, and selection working without it, and he presumably thought Darwin would accept that view:

> Wherefore, so long as gradatory, orderly, and adapted forms in Nature argue design, and at least while the physical cause of variation is utterly unknown and mysterious, we should advise Mr. Darwin to assume, in the philosophy of his hypothesis, that variation has been led [by an intelligent designer] along certain beneficial lines. Streams flowing over a sloping plane by gravitation (here counterpart of natural selection) may have worn their actual channels as they flowed; yet their particular courses may have been assigned; and where we see them forming definite and useful lines of irrigation, after a manner unaccountable on the laws of gravitation and dynamics, we should believe that the distribution was designed (Gray, 1876, 121–2).[10]

William Whewell, whose works on the inductive sciences were among the best-known and most widely read of science treatises in the 1830s and 1840s, had, a good deal earlier, made much the same point, but without reference to Darwin's work:

> Why should the solar year be so long and no longer? Or, this being of such length, why should the vegetable cycle be exactly of the same length? Can this

> be chance?...No chance could produce such a result. And if not by chance, how otherwise could such a coincidence occur, than by an intentional adjustment of these two things to one another? (Whewell, 1833, 28–9).[11]

Whewell's was the standard view in England in the 1830s and 40s. It shows up repeatedly in the nine *Bridgewater Treatises* (one of which was written by Whewell), read and studied by Darwin. Like Herschel before him, Whewell believed God does not intervene personally in the natural plan but had set up "laws" at the creation of the universe, and these have operated throughout geological history, producing species (*CCD* vol. 2, 106; cf. Desmond and Moore [1991], 214; Darwin first referred to Herschel's "mystery of mysteries" in the *Journal of Researches* [second edition, 1845, 27], then again at the beginning of the *Origin* [*Variorum* 71: I. 4]).

INTELLIGENCE

In short, by the time Darwin started contemplating the species question in the late 1830s, respectable scientific opinion had become comfortable with a remote God acting in nature indirectly through designed laws. If Darwin's theory said no more than this—that God ordained the appearance of new species from existing ones through his "magnificent laws" of variation and natural selection, why should anyone be alarmed? The theory would be based on a large body of observable "facts," and yet retain a place for the Creator "in the background."

But, despite Ruse's assertion that Darwin made it "very clear" that he did not mean to refute God's existence in his theory, the evidence from Darwin's writings is more ambiguous. The question of the original source of the laws governing nature is one that Darwin did not frequently address, especially in his published writings. But he did engage in exchanges about the question in the early 1860s with correspondents who raised the question.

One of Darwin's responses was that questions about the ultimate source or origin of natural phenomena, including natural laws, were simply beyond his reach, beyond the reach of science. Just as Newton did not feel a need to explain the source of the law of gravity, but only how it operates, so Darwin saw no need to explain the source of the laws of evolution. He would do enough just to offer an account of what those laws and the processes they govern *are*, so far as they can be discerned.

That sort of answer is a dodge, and it does not say everything Darwin thought. He confessed privately to several of his correspondents that the question of whether God acts in nature, indirectly through "secondary causes" to produce design, was one that he had considered extensively, but always with uncertain results. Good instances of Darwin's private reflections come in his correspondence in the years immediately following the publication of the first edition of *Origin*. Two letters, to Frances Wedgwood and Asa Gray in 1860–1861, are especially revealing. Wedgwood was the oldest daughter of Hensleigh and Frances Wedgwood. She had written an article "The Boundaries of Science. A Second Dialogue," that was published in *Macmillan's Magazine* in July 1861, a copy of which she sent to Darwin. Darwin's reply is as revealing of his private views as any other letter I have seen.

> I think that you understand my book perfectly, and that I find a very rare event with my critics. The ideas in the last page have several times vaguely crossed my mind. Owing to several correspondents I have been led lately to think, or rather to try to think over some of the chief points discussed by you. But the result has been with me a maze—something like thinking on the origin of evil, to which you allude. The mind refuses to look at this universe, being what it is, without having been designed; yet, where one would most expect design, viz. in the structure of a sentient being, the more I think on the subject, the less I can see proof of design. Asa Gray and some others look at each variation, or at least at each beneficial variation (which A. Gray would compare with the rain drops which do not fall on the sea, but on to the land to fertilize it) as having been providentially designed. Yet when I ask him whether he looks at each variation in the rock-pigeon, by which man has made by accumulation a pouter or fantail pigeon, as providentially designed for man's amusement, he does not know what to answer; and if he, or any one, admits [that] these variations are accidental, as far as purpose is concerned (of course not accidental as to their cause or origin); then I can see no reason why he should rank the accumulated variations by which the beautifully adapted woodpecker has been formed, as providentially designed. For it would be easy to imagine the enlarged crop of the pouter, or tail of the fantail, as of some use to birds, in a state of nature, having peculiar habits of life. These are the considerations which perplex me about design; but whether you will care to hear them, I know not (*CCD* 9, 11, 200–1 and nn. 2–8, [July 1861], to Frances Wedgwood).

IS DARWINIAN THEORY "SCIENTIFIC"?

At this point one might worry that any account of evolution that relies upon either "chance" (uncertainty) or God (inscrutability) is an unscientific theory, no better or in principle different, for example, than Paley's "theory" that ascribed life's amazing diversity to divine agency or miracle. Aren't explanations both from "chance" and from "intelligence" equally unverifiable? Or, to put the question somewhat differently, if "falsifiability" is a criterion of any successful scientific theory, can Darwin's theory that evolution by natural selection must rely at some point on chance or divinity be called "scientific"? Is it falsifiable, even in principle?[12]

Elliott Sober has addressed this question in a section of his book *Philosophy of Biology* (1993) by making some important distinctions. The first has to do with the question of which organisms live and die. Sober rightly points out that chance has nothing to do with this. The fitter survive, the less fit die, by an inevitable necessity. Of course, Darwin already occupied all of this territory in the *Origin*.

The point that the fitter are selected to live is conveyed in Sober's chapter section entitled "Why Natural Selection Is Not a Random Process" (Sober, 1993, 36). The title of this section might seem to imply that randomness is not at play *at all* in evolution. But Sober only wants to say that it is not in play in the *second* part of the process—the part wherein natural selection sorts among life's winners and losers on the criterion of fitness. That process is indeed determined and automatic—nothing chancy here at all. This is just another statement of Monod's "necessity," the second half of the title of his book *Chance and Necessity* (discussed in chapter 1 of this volume).

But Sober is clear that randomness must be retained in the *first* part of the process—the generation of variations upon which natural selection works. This is Sober's second distinction. Variations do occur randomly, but in a restricted sense:

> The process of natural selection has two components. First, variation must arise in the population; then, once variation is in place, natural selection can go to work, modifying the frequencies of the variants present. Evolutionists sometimes use the word "random" to describe the mutation process but in a [restricted] sense. Mutations are said to be "random" in that they do not arise because they would be beneficial to the organism in which they occur. There may be physical reasons why a given mutagen—radiation, for example—has a higher probability of producing one mutation than another. "Random mutation" does not mean that the different mutants are equiprobable (Sober, 1993, 37).

This is an echo of Gould's point made earlier: variation is not random in the sense of equally likely in any and all directions, but it is random with respect to whether the variation produced will be well- or ill-suited to the conditions in which it finds itself. Sober summarizes:

> A wind blowing through a junkyard is, near enough, a random process.... But the mutation-selection process differs crucially from [this]. Variation is generated at random, but selection among variants is nonrandom (Sober, 1993, 38).

How, though, does one show scientifically that *any* process is "random," without engaging in question-begging? Sober's answer is in the expression "near enough." A wind blowing through a junkyard is actually *not* random. It is governed by the laws of physics, chemistry, and other natural laws, like all other natural phenomena. Darwin himself made much the same observation in a similar context (*Autobiography*, 1958, 87). But it is "near enough" to random that we probably cannot come much closer to explaining all the variables, forces, and principles at work than by saying "the wind is a random process." Many meteorologists and amateur weather-predictors today would no doubt agree. Consumers of their predictions would agree even more!

"Chance," in other words, in the context of evolutionary biology, is merely a confession of ignorance, as Darwin often said. But in scientific exploration it is no sin to confess ignorance, and often to do so is a virtue, much better than passing the buck to an agency that by definition is inscrutable. Ignorance as a proper explanation is in fact widespread in Darwin's theory. He could not account for the origin of life on earth, but neither did he wish to. That confession does not make the discovery of the origin of life an impossible task, just one Darwin believed he could not resolve successfully. He could not account for every case of geographical distribution, but he had a theory that could account for many cases, and in principle could account for many more (in which belief he has been amply vindicated). He could not explain why some variations came about and others did not, but his theory could, and did, accommodate new developments, such as Mendelian genetics. Above all Darwin could not explain precisely the way variations come about, the causal pathways of variation. He could only make educated guesses. But history, as we have seen, has not been able to improve fundamentally on Darwin's account. We are still left with the brute fact that variations, for all we now know about how they come about, are still beyond full explanatory reach. In that respect they are still "by chance."

"DARWIN'S DANGEROUS IDEA"

The section title is gratefully attributed to Daniel Dennett, who made the phrase the title of an entire book. If there is anything "dangerous" in Darwin's theory, it would seem to be the randomness that exists at the beginning of life and at the beginning of variation. To describe what is dangerous, Dennett has advanced the idea that Darwin's idea of "descent with modification by natural selection" is best understood as an algorithmic process:

> Here, then, is Darwin's dangerous idea: the algorithmic level *is* the level that best accounts for the speed of the antelope, the wing of the eagle, the shape

> of the orchid, the diversity of species, and all the other occasions for wonder in the world of nature. It is hard to believe that something as mindless and mechanical as an algorithm could produce such wonderful things. No matter how impressive the products of an algorithm, the underlying process always consists of nothing but a set of individually mindless steps succeeding each other without intelligent supervision; they are "automatic" by definition: the workings of an automaton. They feed on each other or on blind chance—coin-flips if you like—and nothing else.... Can [evolution] really be the outcome of nothing but a cascade of algorithmic processes feeding on chance? And if so, who designed that cascade? Nobody. It is itself the product of a blind, algorithmic process (Dennett, 1995, 59).

The important properties of an algorithm for Dennett's purposes are that the inputs—the raw materials—of evolution (variations) are "random" and that the outputs are "blindly generated." External Intelligence—God, Mind, Designer—are absolutely unnecessary to explain the generation of the entire Tree of Life; and since there is no other credible evidence that such "skyhooks" have any real existence outside the minds of those who happen to believe they exist, they are dispensable notions in biology, and in science generally. Darwin appears to have had the same thought. Here is Darwin to Lyell in 1859, two months prior to the publication of the *Origin*:

> I would give absolutely nothing for the theory of Natural Selection, if it requires miraculous additions at any one stage of descent.... If I were convinced that I required such additions to the theory of natural selection, I would reject it as rubbish (*CCD* vol. 7, 345).

As usual, Darwin's statement to Lyell does not settle definitively where he stood about intelligence in nature. "Miraculous additions" are obviously ruled out, and if Darwin was consistent in one single belief about his theory, it was this. But this formulation still leaves room for the possibility of divinely crafted "laws" that would make Nature as intelligent as Lyell or many other scientists in Darwin's day could hope for.

A great deal of evidence from Darwin's private writings suggests that he regarded a role for Intelligence in the crafting of the laws of evolution to be as fanciful as the idea that Intelligence directly supervenes in the creation of species or the struggle of organisms in everyday life. Even the *Origin*, in every one of its six editions, suggests Darwin was reluctant or unwilling to bring in Intelligence at all. Yet, at the same time, one finds Darwin through the several editions constantly tinkering with passages in which he utilized the word "chance" (or "accident"), usually trying to disguise its role in subsequent editions. And a careful look at his notebooks from the 1830s and 40s, his "1842 Sketch" and "1844 Essay," and his correspondence, especially during the decade immediately preceding the first publication of the *Origin*, shows that "chance" was a central preoccupation during this entire time of his thinking about species. These private reflections also show that just as he approached the most radical—and dangerous—implications of "chance" in evolution, and stared them in the face, he concluded for whatever reason that it would be best not to include these insights in unvarnished form in his published works. They remained confined for the most part to the relative obscurity of his private notes and confidential confessions to his closest friends until this material was made more widely available through publication in the past 20 years.

Even granting the assumption that Darwin revealed more of his private reflections on sensitive and controversial matters to his closest friends and confidants than to a broader general public, we are still faced with a question about how to interpret the evidence from the letters. It is easy to say that he wavered and equivocated. Each time he wished to remove divine intelligence from the picture of organic evolution, he invariably concluded with a retraction: "I cannot make myself believe everything in the world is just 'by chance.'"

Darwin was deterred from spelling out his deepest convictions in public writings by a host of personal concerns. He had to wrestle with his intellectual inheritance, most of which, dating to his early favorable encounters with the writings of Paley, affirmed a role for divinity. He had to be concerned about his friendships—Lyell above all, but also many other scientists who would view any theory that depended upon randomness in the evolution of life on this planet with extreme skepticism. He had to think about his family—especially his wife Emma—who would no doubt be troubled about the fate of Darwin's own soul were he to abandon God completely. But finally, he had to grapple with his own convictions. These had already been seriously shaken by his encounters with organic nature and, somewhat ironically, with the writings of Charles Lyell. There can be little doubt that he felt ambivalence about the deep issues. He probably often wavered. In fact, it may not matter that much what Darwin believed. But it may nevertheless be useful as a point of historical accuracy to come to the best understanding we can of his own true convictions, especially in light of the fact that his ideas are generally credited with having effected a revolution in science.[13]

But, whatever Darwin thought privately—the reader must draw her own conclusions from the evidence presented—it is clear that he cared a great deal about his public audience and what *they* would think about his views, especially readers who would be encountering Darwin's ideas for the first time in the *Origin*. When he received Wallace's paper in 1858 he knew he would have to act quickly to avoid being scooped in publication in his discovery. He sometimes complained that he could never do full justice to all the "facts" of nature in the "short abstract" that became the *Origin* (a mere 490 pages!), and would try to do so only under the persistent urging of his scientific friends, mainly Lyell. While he claimed he hated to care about "priority," all of the evidence shows that actually he did care, a great deal (Johnson, 2007). But, no sooner had he published, he began to worry again, this time that his theory would be rejected for its "materialism," even implied "atheism." Straightaway he started to make adjustments that would appear in later editions of *Origin*, other later works, and correspondence. The next three chapters show how he tried to diminish the impact of this expected blow as he continued to rework the wording of his theory without changing its substance.

NOTES

1. The usage was preserved and perpetuated by the great comparative anatomist G. Cuvier (1834, I: 1): "In our language, and in most others, the word NATURE is variously employed. At one time it is used to express the qualities a being derives from birth, in opposition to those it may owe to art; at another, the entire mass of beings which compose the universe; and at a third, the laws which govern those beings. It is in this latter sense particularly that we usually personify nature, and through respect, use its name for that of its Creator." Cuvier was considered the "Aristotle of his day," largely because he accepted an Aristotelian account of "nature" (Outram, 1986). Aristotle's influence on Darwin's thought has been explored by Gotthelf (1999); R. Stott (2012); and Johnson (2007).

2. Asa Gray had already issued a similar warning to Darwin in October or November 1857, making the point that "natural selection" suggested a conscious "agent." Darwin had not thought of that before, but confessed in his reply that the term "natural selection" saved him the necessity of spelling out an unconscious natural process in an "incessantly expanded [and] miserably expressed formula" (*CCD* vol. 6, 492). Darwin made the same response to the earliest critics of the *Origin*, in the third and subsequent editions (*Variorum*, 165: 14.6–9*c*).

3. Darwin sometimes utilized expressions in the *Origin* that suggested otherwise: "Man does not actually produce variability; he only unintentionally exposes organic beings to new conditions of life, and then nature acts on the organization, and causes variability" (*Origin* [*Variorum*], 730). He let this (and similar) statement stand through all editions. But, as I will show, this was for Darwin a mere shorthand for what he knew to be a far more complex process—or rather, an easy way to duck the major issue in variation. To say "nature causes variation" is, he knew, no more illuminating by itself than to say "the Creator" causes it or a "vital force" causes it.

4. I owe this observation to J. Beatty (2006), who in turn cites Rothschild and Jordan (1903). The contemporary source is Nilsson (1998).

5. "Astronomers might formerly have said that God ordered each planet to move in its particular destiny.—In the same manner God orders each animal created with certain form in certain country, but how much more simple, & sublime let attraction [of gravity] to act according

to certain laws such are inevitable conseque[nce] let animal be created, then by the fixed laws of generation, such will be their successors" (*CDN* B-101). Other naturalists prior to Darwin had argued for much the same view, notably F. Cuvier and Étienne Geoffroy St. Hilaire. Parkes's *Chemical Catechism* (1818) and Whewell's *Bridgewater Treatise* (1833) both emphasized "divinely established general laws" as the best explanation for natural phenomena. Darwin acknowledged such views on this subject, without fully endorsing them, in the B Notebook: cf. *CDN* B-111–15. Cf. Schweber (1985, 42, in D. Kohn).

6. Darwin again famously referred to life being "breathed into" living creatures in the final sentence of the *Origin*, often quoted. In the same sentence he compared the continuous production of new species through time to the continuous "cycling of planets according to the fixed law of gravity," implying at best that the creative force, if there is one, has left the governance of the earth to the operation of fixed laws. Somewhat strangely, Darwin changed this last sentence of the *Origin* in the second edition (perhaps prodded by his wife Emma, a suggestion made to me by an anonymous reviewer of the manuscript) to read "breathed by the Creator into a few forms or into one" (*Origin* [*Variorum*], 759: XIV. 270). He later regretted making this change as an unintended capitulation to the "Pentateuchal" point of view.

7. Ruse has strengthened his stance in more recent work, in his chapter to the co-edited volume *The Cambridge Companion to the "Origin of Species"* (Ruse and Richards, 2009, 2): "In religion, [Darwin's trip in Wales with Adam Sedgwick in 1831] was important because Darwin's rather literalistic Christianity started to fade and he became something of a deist, believing in God as unmoved mover and that the greatest signs of His powers are the workings of unbroken law rather than signs of miraculous intervention." See also Ruse and Richards (8) for evidence that suggests they believe Darwin remained a deist from then on.

8. Darwin encountered Herschel's comment in the second edition of Babbage's *The Ninth Bridgewater Treatise* (London, 1838), who was quoting from a letter from Herschel to Charles Lyell, February 20, 1836. Cf. *CDN* E-59 and n. 2.

9. Darwin himself later confessed that he regretted inserting a reference to creation in his work, claiming that what he really meant was that species "'appeared' by some wholly unknown process" (*LLD*, 3:18).

10. Darwin did much to encourage Gray (and no doubt many others) to believe that he (Darwin) really did accept Gray's explanation. In several of his letters to Gray after Gray's first comments on Darwin's theory, Darwin emphasized how well Gray had understood him. He even went to some lengths to ensure that Gray's views received a wide hearing in England, especially by promoting and financing the publication of a pamphlet that assembled some of Gray's writings on the subject in a small, separate volume.

11. Whewell, in fact, despite his commitment to induction in science, could never give up a role for miracles in organic adaptations. Even natural laws could not do all the work necessary to explain organic adaptations: "Nothing has been pointed out in the existing order of things which has any analogy or resemblance, of any valid kind, to that creative energy which must be exerted in the creation of new species" (1840, 2: 133–4). Darwin, as much as he may have followed Whewell in other ways, could not go along with him here.

12. The principle of falsifiability as a criterion of a good scientific theory was introduced into the philosophy of science by Karl Popper in his 1936 *Logic of Scientific Discovery.* On this view, for example, Freud's theory of an Oedipus complex cannot be falsified, for whether one wants to kill one's father ("proof" of an aggressive instinct) or one does not ("proof" of a repressive

psyche), the theory is validated. No possible falsification is available, and that renders a theory unscientific. For a time Popper himself, while having high regard for Darwin's theory, nevertheless did not believe it is falsifiable (*Unended Quest; An Intellectual Autobiography*, 1976, ch. 37). Late in life he somewhat changed his opinion, showing that at least some of the propositions of the theory could be tested scientifically (in Radnitsky and Barlow, 1993, 144–145).

13. Whatever Darwin may have believed in private or said in public, his revolution is generally believed to have removed "design" from natural processes. Alexander Rosenberg (1985, 246) sums up this way: "One of the most salient features of Darwin's revolution is that it ended forever the biological appeal of the argument from design, which founded the teleology of nature on the desires, intentions, and conscious designs of God."

5

Darwin's Evolving Views about Chance

FROM CHANCE TRANSPORT TO CHANCE VARIATION: *NOTEBOOKS*

The idea of chance transport obviated the need for a theory of special independent creation for many creatures, but it also brought forward a new question: from what causes do these myriad special abilities for travel arise? That question brought Darwin head-to-head with what would become his new line of thinking in the *Notebooks* and thereafter: the question of how new variations arise, and how their appearance may play a role in evolution. They too may be the result of what may be called "chance." But that discovery soon posed a quite different challenge to

Darwin: how to present his theory in a scientifically respectable way. Predictably, many readers would not take a liking to any theory whose very foundation was chance variation.

Darwin's discovery of "chance variation" should be dated to the earliest *Notebooks* on transmutation, *Notebooks* B and C.[1] It may plausibly be taken to coincide in time with his revelation that he had a "theory of his own," especially as it contrasted with Lamarck's theory. "My theory very different from Lamarcks [*sic*]," he observed to himself either in late 1837 or early 1838 (*CDN* B-214). This was long before he read Malthus. He registered his earliest thoughts on chance variation in *Notebook* B, with an emphasis that would henceforth accompany many of his reflections on this subject:

> Why should we have in open country a ground woodpecker <<ditto with parrot>>? A desert Kingfisher? Mountain tringas? Upland goose? Water chinois [or] water rat with land structures?:*Law of chance would cause this to have happened in all, but less in water birds*—carrion eagles [etc.] (*CDN* B-55e, emphasis in the original).

A "law of chance"? That is a curious expression, because "law" and "chance" would seem to be opposed to one another. Darwin would soon come to recognize the incompatibility, but he would never—even in later life—come to resolve the inconsistency, at least in public expression, in favor of one or the other. Both would be retained throughout his theories of change as playing important roles. But this early entry does show that Darwin was up against the idea, even at this early stage, that some of the weirdest adaptations in nature must be haphazard, without rhyme or reason. A woodpecker, for example, so plainly made for climbing and pecking in trees, might be found where no trees exist, and yet still be able to survive. Only "chance" could explain such strange variations. As he remarked in *Notebook* C, "In [the] round of chances every family will have some aberrant groups" (*CDN* C-73). That is just the way Nature is.

For a time Darwin seemed uncomfortable with the idea that Nature is just "by chance." Alongside chance he entertained the thought that "Nature" is a conscious, intelligent agent, one that makes plans to ensure survival and stability:

> The infertility of cross & cross, is method of nature to prevent the picking of monstrosities as man does.—One is tempted to exclaim that nature [is] conscious of the principle of incessant change in her offspring [and] has invented all kinds of plan to insure stability (*CDN* C-52–3).

And again, some pages later:

> Even a deformity may be looked at as the best attempt of nature under certain very unfavorable conditions [to ensure survival] (*CDN* C-65).

Late in *Notebook* C Darwin was still affirming a role for an intelligent Nature, this time arguing that "nature" can produce what she needs to produce to ensure viability of species:

> The females of some moths, like glowworms have <<these abortive organs in some Male animals, Mammae in Men, capable of giving milk>> rudimentary wings. So nature can produce in sex, what she does in species of Apterix (*CDN* C-215e).

But by this time, mid-1838, Darwin began to doubt that Nature is intelligent, at least insofar as he felt comfortable assigning it a role in causing variations to occur. Increasingly, instead, he relied on the idea of "chance." His first reflection along these lines—chance variation with no assignable reason in intelligence—appears to come midway through *Notebook* C:

> Aberrant groups...causes the confusion in this system of nature. Whether [a] species may not be made by a little more vigour being given to the chance offspring who have any slight peculiarity of structure. [Several examples are given] (*CDN* C-61).

Another instance is given late in C:

> Dwight's Travels in America, speaks of short legged sheep. Hereditary proceeding from an accident. New England farmer,—useful, could not leap fences (*CDN* C-221e).

In the foregoing cases it is apparent that Darwin had touched on the idea that would henceforth govern his thinking about how selection works, both under domestication and, by inference, in nature. Nature gives up an "accidental" variation, that variation is sometimes "useful" (either to the breeder for animals under domestication or for the organism itself for its survival in nature), if it is "selected," and having been selected, its variation is handed down to progeny (i.e., is hereditary). Before he ever read Malthus, Darwin had discovered his "dangerous idea" about how speciation occurs.

But he took a detour from this line of thinking through most of *Notebook* D. He did not abandon chance as an explanation for variation, but his interests shifted. In particular, he began to ask about the importance of "laws" in accounts of variation. The aim of science, he came to believe, is to reduce "facts" to "laws" in order to arrive at "predictions" rather than mere explanations (*CDN* D-67–9). He also starts in earnest to speculate about what the "laws" governing variation might be. For example, having been "struck looking at the Indian cattle with Bump, together with Bison" and noting the resemblance, Darwin wondered to himself if "there [is] any law of this" (*CDN* D-65). He concluded tentatively in the affirmative, but was also cautious in drawing any firm conclusions:

> Is there some law in nature an animal may acquire organs, but lose them with more difficulty... hence become EXTINCT, & hence the IMPROVEMENTS of every type of organization. Such [a] law would explain everything.—PURE HYPOTHESIS be careful (*CDN* D-58, emphasis in original).

But if variation *is* governed by law, what becomes of chance? Darwin wondered the same thing later in *Notebook* D:

> Is there any *law of variation* <<(as Hunter supposes with Monsters)>>[?] If armless cat can propagate, ie with chance of two being born at same time, & make breed, one would doubt any law.—Yet seeing the feathers along one toe of the Pouter one thinks there is a law—that there must have been a tendency for feathers to grow there (*CDN* D-112).

Chance, or law? In light of the ambiguous evidence for answering this question and the difficulty of interpreting it, Darwin could not decide. In *Notebook* E, while he retained an important role for "law" (*CDN* E-51–3), he returned to "chance" as at least one acceptable explanation for change. To Kant's assertion that "nothing [in nature] happens by chance" (where Kant was referring expressly to the "assumption" that no organ or structure in any creature is "in vain"), Darwin retorted: "All this reasoning is vitiated when we look at animals on my view" (in *CDN*, 412, n. 57-4, quoting Darwin's marginal comments to Whewell [1837]; the editors of *CDN* note that Darwin had read this book by the time of *Notebook* E, i.e., late 1838 to early 1839). What Darwin meant was that some structures and organs have no discernable adaptive value whatsoever and still persist, which can only mean that nature does in fact come up with "chance" variations after all. "Chance" may not be the only factor at work in change, but it is one such factor.

Darwin stuck with this idea throughout the rest of the *Notebooks*. For example, later in E he registered disagreement with William Herbert's assertion that "Plants do not become acclimatized by crossing or by *accidental production of seedling with hardier constitution*" (*CDN* E-111, Darwin's emphasis). Darwin's view was quite the contrary:

> My principle [is] the destruction of all the less hardy [seedlings], & the <*accidental*> preservation of <<*accidental*>> hardy seedlings (which are confessed by Herbert) to sift out the weaker ones: there ought to be no weeding, but a vigorous battle between strong and weak (*CDN* E-111–12).

The fact that Darwin underlined the word "accidental" in both places suggests that he attached unusual importance to it. And the fact that he changed the word's position in the sentence shows this was no casual or passing thought. The first placement, as modifying "preservation," would just be another instance of the idea, encountered often before, of "better chances of survival" of some organisms over others. Darwin did not mean that here, and he was clear that he did not mean it. He meant that the "hardy seedlings" were "by chance," that they had come about as the "raw material" of natural selection by a process he could only describe as accidental. By the time of *Notebook* E Darwin had settled into the conviction that "chance" played a role in evolution that he could no longer dismiss or overlook. The conviction only deepened in the 1838–1839 "metaphysical notebooks" M and N (Kohn and Hodge, in Kohn, 1985).

DARWIN'S ABSTRACT OF MACCULLOCH

Further evidence of Darwin's shifting interest in late 1838 toward the role of chance in variation emerges in an abstract Darwin made of John Macculloch's 1837 work *Proofs and Illustrations of the Attributes of God*. The editors of the *Notebooks* draw attention to Darwin's "bold and assertive references to Malthus... along with a critical stance toward natural theology" (*CDN*, 631). What is equally noteworthy is Darwin's ridiculing tone of Macculloch's account of organisms displaying "special adaptations" that, Macculloch believed, have obviously been designed with an eye toward their usefulness, in other words, of any naturalistic teleology. His view is that many "adaptations" are best understood as purely accidental variations:

> It is hard on my theory of gain of small advantages thus to explain the curling of the valves of the broom, or the springing of other seeds. But are we certain

> that these are necessary adaptations? May they not be accidental? We have good reason to know that they would not be detrimental accidents, & domesticated variations show us accidents may become hereditary [produce some peculiarity in seed vessel] if man takes care they are not detrimental (Darwin referring to Macculloch, 53v).

That the word "accidents" (or "accidental") appears three times in this short paragraph should be enough to convince us of Darwin's skepticism toward explanations, like Macculloch's, that trace change to some intentional purpose or design, put into plants by some unnamed power to provide them against destruction and to suit them for survival in difficult environmental circumstances. But Darwin does not hesitate to close the noose on such explanations.

> Mac. has long rigamarole [*sic*] about plants being created to arrest mud &c at deltas. Now my theory makes all organic beings perfectly adapted to all situations, where in accordance with certain laws they can live.—Hence the mistake they are created for them. If we once venture to say plants created to prevent the valuable soil in its seaward course, we sink into such contemptible queries, as why the earth should have drifted; why should plants require earth, why not created to live on alpine pinnacle? If we once to presume that God <<created plants>> to arrest earth (like a Dutchman plants them to stop the moving sand) we <<do>> lower the creator to the standard of one of his weak creations (Darwin referring to Macculloch, 54r).

The same antiteleological position is brought forward again late in the same *Notebook*, again in direct challenge to Macculloch's arguments for a "final cause" discernable in all natural productions. Darwin adds his own counter immediately after referring to Macculloch:

> This kind of doctrine runs through Macculloch, the bills of the Grallae <<have been made>> long <<as adapted to>> because their food lies deep.—I say it is <<as>> simple consequence they become long, not at once, but by steps, of which we have manifold traces in the several genera of Grallae (Darwin referring to Macculloch, 28v).

The proper explanation for variation is not design, but chance:

> Suppose six puppies are born, <<& it so chances that one out of every hundred litters is born with long legs>> & in the Malthusian rush for life, only

> two of them live to breed, if circumstances determine that the long legged one shall rather oftener than any other one survive. In ten thousand years the long legged race will get the upper hand (Darwin referring to Macculloch, 28v).

The implications of Darwin's account of chance for a natural teleology are then fully spelled out, in a passage reminiscent of a passage that Darwin transcribed into the "Historical Sketch" of the *Origin* from his translated copy of Aristotle's *Physics:*

> All such facts are merely relations of one general law. The plants were no more created to arrest earth, than the earth revolves to form rain to wash down earth from the mountains upheaved by volcanic force (Darwin referring to Macculloch, 54r).[2] *What bosch!!* Darwin exclaimed (Macculloch, 54v, Darwin's emphasis).

Part of Darwin's evident contempt for Macculloch's arguments stemmed from his early realization that many "adaptive structures" are not readily understood as adaptations at all. Instead they often seem to be useless or even counterproductive structures that nevertheless do not condemn organisms to certain extinction. The flip side of this coin is that some structures that seem adaptive (i.e., useful for survival) do *not* result in preservation of the organism or its species. In other words, close inspection reveals in nature something far different from the preferred account of the teleologist and natural theologian that "there is a place for everything, and everything has its place." Some structures that endure are non-adaptive for survival, and other structures poorly suited to survival endure anyway. For example, to Macculloch's argument that a camel's stomach "offers another of those special contrivances... so perfectly adapted, that the design has been universally admitted," Darwin registered disagreement, finding this case to exhibit not obvious design at all but rather a "puzzle" (Darwin referring to Macculloch, 57v). Darwin also objected emphatically to Macculloch's opinion that the lack of a proboscis in certain insects would have produced an "inconvenience" for them:

> [Macculloch] says *inconvenience* would have arisen had <<some>> insects not been provided <<with proboscis, as bee and butterfly>>. Inconvenience! *Extinction*, utter *extinction*! Let him study Malthus and Decandolle (Darwin referring to Macculloch, 57v, original emphasis).

Once Darwin started down this line of reasoning, his writing reached a stronger emotional pitch. The entire edifice of teleological design in Nature had to be dismantled, including even Darwin's own habit of using the term "final cause" in his writing":

> The Final cause of innumerable eggs is explained by Malthus. <<It is anomaly in me to talk of Final causes. Consider this!>> Consider these barren Virgins. Macculloch, 235, talks of the long spinous processes in Giraffe &c, as adaptations to long necks. He may as well say <<long>> neck is adapted to long necks, 236. Marsupial bones especial adaptation to <<young>>. Good God, & yet Males have them. What trash (Darwin referring to Macculloch, 58r).

What ensures the survival of any new form is that one small change, arising by chance, is better suited to conditions than 10,000 other small changes. No surviving change needs to be "perfect," only non-fatal and better for survival than other changes:

> I look at every adaptation, as the surviving one of ten thousand trials [chances?]. Each step being perfect <<or nearly so, although having hereditary superfluities. Man could exist without Mammae>> to the then existing conditions (Darwin referring to Macculloch, 58v).[3]

THE "SKETCH" OF 1842

In 1842, three years after Darwin had finished making entries in his transmutationist notebooks he wrote a brief "Sketch" of his overall theory, not intended for publication. He expanded the Sketch two years later in an "Essay" that he believed would suffice for publication if he should suddenly die, although he did not write it for publication either.[4] Both documents are interesting for showing the state of Darwin's private thinking on the full range of questions he would later take up in the *Origin* and also for enabling us to track changes and adjustments in his language and thought as he went along. Both documents do capture the essence of his theory—variation, heredity, superfecundity, natural selection, all acting together to form new races and species through geological time. What do the documents show about how Darwin understood variation?

Darwin was still not sure in 1842 about the causes of variation. He wavered among three different accounts: divine agency acting indirectly through laws; "Lamarckian" and "Geoffroyian" factors of the direct action of physical conditions on habits and organization; and chance. He had certainly not yet *excluded* a role for divine agency. On this scheme humans are to be seen as the apex of evolutionary development, the deliberate goal set for nature by an omnipotent God. Near the conclusion of the "Sketch" he wrote:

> It accords with what we know of the law impressed on matter by the Creator, that the creation and extinction of forms, like the birth and death of individuals should be the effect of secondary [laws] means.... From death, famine, rapine, and the concealed war of nature we can see that the highest good, which we can conceive, the creation of the higher animals has directly come. Doubtless it at first transcends our humble powers to conceive laws capable of creating individual organisms, each characterized by the most exquisite workmanship and widely extended adaptations. It accords better with [our modesty] the lowness of our faculties to suppose each must require the fiat of a creator, but in the same proportion the existence of such laws should exalt our notion of the power of an omniscient Creator (Darwin and Wallace, 1858, in *EBNS* 86–7).

The idea here is that an intelligent God created laws through the operation of which He would effect his divine plan—an orderly unfolding of natural forms culminating in the creation of "the higher animals," and that the plan itself is the "highest good." Such a formulation could hardly have proven objectionable to anyone who had already accepted the fact of transmutation, no matter what his religious beliefs. Darwin was certainly not removing intelligence from the natural scheme; he was only removing the divinity from an active and daily role in shaping the structures populating the earth. As he had already said much earlier in the "Sketch":

> Who, seeing how plants vary in the garden, what blind foolish man has done in a few years, will deny an all-seeing being [by which Darwin here means "Natural Selection"] in thousands of years could effect (if the Creator chose to do so), either by his own direct foresight or by intermediate means—which will represent the creator of the universe (45–6).

The "intermediate means" Darwin has in mind here is "natural selection." Here it is cast as a being "infinitely more sagacious than man," endowed with foresight and perception, and, as such, able to "select" favorable variations for survival and unfavorable ones for destruction with unerring accuracy. But natural selection is clearly distinguished from an "omniscient Creator":

> But if ever part of a plant or animal was to vary... and if a being infinitely more sagacious than man (not an omniscient creator) during thousands and thousands of years were to select all [favorable] variations (45–6).

In other words, at this point Darwin has carved out a role in his theory for two distinct intelligences: "natural selection," which is endowed with wisdom and foresight, but not omniscience, and an "omnipotent Creator" who has fashioned "Natural Selection" as the "intermediate means" (i.e., laws) through which He wills his plan for the earth.

By the time Darwin wrote the *Origin*, or at least by his later revisions of it, he was explicitly disavowing any role for intelligence in "nature," claiming such expressions as implied intelligence were purely "metaphorical," necessary for "brevity's sake" (*Origin* [*Variorum*], 164–5: IV. 14.1–9: *c*). He did not go that far in the 1842 "Sketch." He did, however, identify what he took the laws of variation to be. Mainly they are the Geoffroyian principle of the direct action of external conditions on organization (e.g., a change in diet causing changes in body size: 44); the Lamarckian principle of the direct action of external conditions on habits, with the change in habits in turn effecting changes in organization (e.g., a giraffe stretching its neck to reach for higher food: 55); and external conditions acting on the reproductive system (42). Changes thus acquired are then said to be passed down to offspring.

Where, then, is "chance"? After describing the operation of what he later came to call "sexual selection" in animals, Darwin wrote, seemingly as a reminder to himself: "Introduce here contrast with Lamarck—absurdity of habit, or chance?? or external conditions, making a woodpecker adapted to a tree" (49).

Darwin's son Sir Francis Darwin, who first edited and published the "Sketch" in 1909, remarked about this passage that "It is not obvious why the author objects to 'chance' or 'external conditions making a woodpecker,'" adding that both factors (i.e., habit and chance) had earlier been allowed a role in variation (49, n. 1). But Darwin's objection to "chance or external conditions making a woodpecker" can perhaps be explained by taking a closer look in the "Sketch" at how Darwin understood the operation of external conditions in general.

When Darwin first introduced "external conditions" as a cause of change he stipulated two different kinds of this agency: (1) conditions acting directly on organisms, causing them to change either their structures (e.g., cold climate causes longer hair) or their habits (e.g., giraffes having to stretch to reach higher leaves); (2) changed conditions cause variation "not [as] direct effect of external conditions, but only in as much as it affects the reproductive functions" (42). This latter agency Darwin later came to call "indirect effects of external conditions," and it had a special fascination for him. From an early time he was intrigued by the finding of plant and animal breeders that "wild animals [and plants], taken out of their natural conditions, seldom breed" (50–1), and he appears to have become convinced from an early time—certainly by the time of the "Sketch"—that this fact could only be explained by conditions somehow working on the reproductive system *prior to* conception (41, 50, 52–3).[5] Thus, in explaining why animals taken from their natural conditions and placed in captivity often do not breed Darwin surmised that changed conditions must have brought about "constitutional differences" in the animals, a phrase which in the *Origin* was explicitly equated with the reproductive system (50). The same, he argued, was true with plants that are exposed to new conditions (51). Summing up he wrote:

> We see throughout [the plant and animal kingdoms] a connexion between the reproductive faculties and exposure to changed conditions of life whether by crossing or exposure of the individuals (53).

The identification of the reproductive organs as an important source of variation was a significant breakthrough for Darwin, constituting one of the major differences between his theory and Lamarck's, and accounting no doubt for his desire to contrast his views with the latter's. At the same time, to say that conditions affect reproductive organs was not a complete answer even about this source of variation. How does the environment work on reproductive organs exactly, and what changes would thereby be induced? In a way, Darwin's new answer was no more adequate than Lamarck's "response to felt needs" and his "interior sentiment." Both answers specified an agency of change in such a way as simply to raise new questions. That was Darwin's complaint against Lamarck, but what can he say against the same charge if it is brought against him?

This is the point at which Darwin turned to the notion of "chance," and I believe he turned to it in direct response to the need to answer the potential criticism that he had not yet solved the problem of variation.

Darwin's first mention of chance seems to suggest that variations producing structures wonderfully adapted to diverse ends cannot be the result of chance: "Can varieties be produced adapted to end, which cannot possibly influence their structure and which it is absurd to look at as effects of chance" (44–5). But the passage may be read as suggesting that, while some variations cannot be due to chance, *some other* variations *are* due to chance:

> [Some] variations [are] the direct and necessary effects of causes, which we can see act on them, as size of body from amount of food, effect of certain kinds of food on certain parts of bodies, etc.; such new varieties may then become adapted to those external [natural] agencies which act on them. But can varieties be produced adapted to end... which it is absurd to look at as effect of chance" (44–5).

The passage is difficult, but it could mean that some variations, those in which conditions work directly on organisms *are* by chance, whereas some variations evidently cannot be. Which are of the latter sort? This question comes up at just the point in the text where Darwin introduced his extended metaphor of natural selection as an "infinitely wise being." Thus, it is plausible to take him as meaning that this very wise being sometimes actually "foresees" what new organs for survival will be needed by plants and animals in an ever-changing environment and expressly makes provisions to ensure that they will be equipped with them. Since the whole process in these cases is governed by intelligence and foresight, it is obviously "absurd" to look at *these* adaptations as the effect of chance. But that leaves other adaptations, where conditions *are seen* to act on organisms, especially when they act on reproductive functions, to be attributable to chance. In other words, when external conditions rather than intelligence governs change any result is possible, and so we can say "by chance." When intelligence governs change, the expression "by chance" is "absurd."

If this reading is correct, it leads to a startling conclusion. We know that Darwin eventually abandoned the notion that natural selection acts by foresight and wisdom, substituting instead the idea that it is a blind and mechanical process that weeds out the unfit and preserves the fit "automatically," once variations are given. But if an intelligent Nature thus ceases to be responsible for variation in any sense, that leaves only "chance" as ultimately responsible. One may still speak of physical conditions working on organs and organisms, thus causing them to change. But such expressions as these—quite common in the "Sketch"—really boil down to the idea that we cannot know precisely how this process of change works, its inner secret, as it were. About such changes the only thing we can say is they are "by chance."

That reading comports well with what Darwin says later in the "Sketch."[6] In Part II Darwin on several occasions alluded to the origins of variation, not as a systematic inquiry, as in Part I, but in passing while discussing other questions. In these instances he usually refers to the origin of new forms as due to "chance." Speaking of affinities and classification, for example, he wrote:

> According to mere chance every existing species may generate another, but if any species A, in changing get an advantage and that advantage (whatever it may be, intellect, etc., or some particular structure or constitution) is inherited, A will be the progenitor of several genera or even families in the hard struggle of nature (74).

Or again, speaking of the retention by some organisms of apparently useless structures, he wrote:

> Now according to our theory during the infinite number of changes, we might expect that an organ used for a purpose might be used for a different one by his

> descendant.... And if it so chanced that traces of the former use and structure of the part should be retained [its continued existence] instead of being utterly unintelligible becomes simple matter of fact (77).

And then, in connection with the existence of abortive organs, Darwin observed:

> The term abortive organ has been thus applied to above structure... from their absolute similarity to monstrous cases, where from *accident*, certain organs are not developed.... Nectaries abort into petals in columbine *aquilegia*, produced from some accident and then become hereditary... These cases have been produced suddenly by accident in early growth... (82, emphasis in original).

Admittedly, Darwin was still working out some aspects of his theory in 1842 and was still trying to decide how to put the various pieces of it together into an effective demonstration. One of these pieces was the question of the causes of variation—a question Darwin continued to ponder without complete resolution all the way through the final edition of the *Origin*. What we find in the "Sketch" is an early encounter with the possibility that, while some variation appears directed toward adaptive needs of organisms through the deliberate foresight of an intelligent "nature," at least some variation can be assigned no better cause than "chance."

THE "ESSAY" OF 1844

The "Essay" of 1844 makes an advance over the 1842 "Sketch" on the question of variation, but without substantially altering the main argument. The "syllogistic core" of Darwin's theory, presented at the end of Chapter 2 of the "Essay," is essentially the same as that found in the 1842 "Sketch." Variation is again presented as the first step in the sequence of the argument, and variations are said to be "given by nature" (99). Causes of variations are of two kinds, direct and indirect, a distinction that would be preserved and expanded in the *Origin*. Direct causes are those that operate on adult living organisms, inducing change in form or in habits, and the changes are transmitted hereditarily to subsequent generations. Indirect causes are those induced earlier in the life of an organism—either prior to conception (i.e., action on the reproductive organs of the parents), during the act of conception (not further explained), or during the embryonic stage (95). In addition, variation is now shown to be limited in degree and kind by the nature of the organisms, the nature of the conditions, the tendency to reversion, and the "laws" of interconnected variation, later to be called "laws of correlations of growth" (105–7).

Of the two sources of change mentioned above, indirect influences are, as in the "Sketch," much more important than direct influences; and further, of the various ways indirect influences can work, their effect on the "reproductive system" is the most important. Speaking of variation of organisms under domestication, Darwin put the idea this way:

> We must attribute [the great degree of variation among organic beings under domestication] to the indirect effects of domestication on the action of the reproductive system.... Judging from the vast number of new varieties of plants which have been produced in the same district under nearly the same routine of culture, that probably the indirect effects of domestication [on the reproductive system] in making the organization plastic, is a much more efficient source of variation than any direct effect which external causes may have on the colour, texture, or form of each part (95–6; Schweber, 1985, 56–7 in Kohn, 1985).

What is true of plants and animals under domestication is assumed to be true also of plants and animals in a natural state. This is because geology informs us that physical conditions of the earth are constantly changing, even if very slowly. If variations among organisms in domestication are induced by the new physical conditions in which these organisms are placed, it must follow that changes in physical conditions in nature will also induce changes:

> Domestication seems to resolve itself into a change from natural conditions of the species...; if this be so, organisms in a state of nature must *occasionally*, in the course of ages, be exposed to analogous influences; for geology clearly shows that many places must, in the course of time, become exposed to the widest range of climate and other influences.... Whatever might be the result of these slow geological changes, we may feel sure... that organisms must suddenly be introduced into new regions, where, if the conditions of existence are not so foreign as to cause its extermination, it will often be propagated under circumstances still more closely analogous to those of domestication; and therefore we expect will evince a tendency to vary (113).

And, as is true of variation among domesticated organisms, variation among organisms in nature can usually be traced to conditions working on the reproductive system:

> The [direct] effect [of new conditions on organisms] we might expect would influence in some small degree the size, colour, nature of covering, etc., and

> from inexplicable influences even special parts and organs of the body. But we might further (and this is far more important) expect that the reproductive system would be affected, as under domesticity, and the structure of the offspring rendered in some degree plastic (114).

Conditions working on the reproductive system continue to be singled out as the main source of variation throughout the "Essay."[7] It is also important to note that in nearly every instance where Darwin points to this factor as the "most efficient" in causing variation, he juxtaposes it against the Geoffroyian/Lamarckian idea of "direct effects of conditions on existing organisms," as in the passage quoted directly above. In other words, by the time of the 1844 "Essay," Darwin had become much clearer that the mystery of variation was better sought in conditions working indirectly on reproductive organs than in conditions working directly on existing organisms. And he was much less shy about saying so than he eventually became during later editions of the *Origin*.

But are such variations "by chance," that is, blind with respect to future adaptive needs? Darwin's position in the "Essay" is again somewhat murky. In some sense change in nature, in the sense of the creation of new races and species, cannot be "by chance" if Darwin intended his account of natural selection to be taken literally rather than metaphorically. For here again, as in the "Sketch," natural selection is presented as an extremely intelligent Being that "forms new races" by looking ahead to future adaptive needs and then supplying organisms with organs and habits that are necessary for survival. Immediately after explaining that mere variation without selection would seldom produce any lasting change among organisms, Darwin wrote:

> Let us now suppose a Being with penetration sufficient to perceive differences in the outer and innermost organization quite imperceptible to man, and with forethought extending over future centuries to watch with unerring care and select for any object the offspring of an organism produced under the foregoing circumstances [of a constantly changing physical world], I can see no conceivable reason why he could not form a new race, or several... adapted to new ends. As we assume his discrimination, and his forethought, and his steadiness of object, to be incomparably greater than those qualities in man, so we may suppose the beauty and complications of the adaptations of the new races and their differences from the original stock to be greater than in the domestic races produced by man's agency (114–5).

Natural selection, then, is still at this point an intelligent being busily at work in forming new races according to his "discrimination and foresight." But this intelligent being is not the Creator. It is, rather, the creation of the Creator, as is everything in creation:

> He will be a bold person who will positively put limits to what the supposed Being [viz., Natural Selection] could effect during whole geological periods. In accordance with the plan by which this universe seems governed by the Creator, let us consider whether there exist any *secondary* means in the economy of nature by which the process of selection could go on adapting organisms... to diverse ends. I believe such secondary means to exist (116).

This passage is a clear echo of that found in the "Sketch," at just the point where Darwin had declared "absurd" the idea that variations are "by chance." It shows that Darwin continued to entertain the possibility that two intelligences exist in Creation, the first a Creator who creates "secondary means" (i.e., "laws of nature") according to which he governs the universe: "It is in every case more conformable with what we know of the government of this earth, that the Creator should have imposed only general laws [than that He should have ruled through an infinite series of individual acts]" (154; cf. also 253); and second, an "Intelligent Selector" who makes choices about how to fit out favored races with the right equipment to survive and about what individuals and races must perish for want of such equipment. But, as we saw in the "Sketch," Darwin's account of natural selection as an intelligent being is not incompatible with a role for chance in variation. Darwin was committed to such a role in the "Sketch." Did he remain committed to that view in the "Essay"?

The answer is less than obvious. In the "Essay" we find frequent mention of "unknown agents" or "unknown causes" determining variation. In one sense, if variation is "caused," even if the causes are unknown, then it cannot be due to "chance." A "cause" implies a regular, knowable, and predictable sequence, whereas "chance" implies unknowability even in principle. On the other hand, Darwin has as yet only explained *that* conditions work on organisms or on their reproductive organs so as to effect change, not precisely *how*. The mechanism of change has not been spelled out in detail. Natural selection as an intelligent being exercising foresight about which organisms to preserve and which to destroy does not solve this problem. As Darwin said, variations are "given by nature" and only then "selected" by natural selection. But does nature exercise intelligence when it "gives" "variations," or does nature simply give them randomly?

The answer is that nature gives changes randomly. Darwin does not spell this out as plainly as one looking for chance in his theory might wish, but the idea is there that "chance" is an irreducible fact in variation—and not only in variation. For example, "chance" is mentioned in the "Essay" in connection with the fortuity of appearance of existing species in new regions where they have "chanced" to arrive (e.g., 114, 202), and more often, in connection with the "better chance of survival" of organisms better equipped to survive in changing conditions than others (e.g., 119, 148, 153, 198, 242). In both of these contexts the use of the idea of "chance" implies an element of the accidental in how natural selection works—viz., not according to a plan at all, but according to blind fortune.

But the message of fortuity is most decisively driven home in Darwin's various discussions of how we are to understand *why* some variations are more likely to come about than others. Even if we know that conditions affect organisms both directly and indirectly, it is hard to get away from reducing these influences to an unknowable chance. Breeders of domesticated animals and plants, for example, select for "chance domestic variations," that is, variations that come about for no assignable reasons (222). In discussing nature, where the question of the source of the "raw material" on which the "intelligent being" called natural selection could act, Darwin commented as follows:

> For instance, let this imaginary Being wish, from seeing a plant growing on the decaying matter in a forest and choked by other plants, to give it power of growing on the stems of rotten trees, he would commence selecting every seedling whose berries were in the smallest degree more attractive to tree-frequenting birds... He might then, if the organization of the plant was plastic, attempt by continued selection of chance seedlings to make it grow on less and less rotten wood, till it would grow on sound wood (115).

Again, speaking of instincts among animals, Darwin was quite sure that these too were subject to variability, but he was not sure how such variations arise. Thus, he wrote:

> Knowing that a dog has been taught to point, one would suppose that this quality in pointer-dogs was the simple result of habit, but some facts, with respect to the occasional appearance of a similar quality in other dogs, would make one suspect that it originally appeared in a less perfect degree, *"by chance,"* that is from a congenital tendency in the parent of the breed of pointers (139, emphasis in the original).

"Congenital influences" are mentioned again later in the same chapter on instincts, in a passage immediately followed by the following:

> Now suppose this bird [exhibiting a novel instinct about where to lay its eggs] to range slowly into a climate which was cooler, and where leaves were more

> abundant, in that case, those individuals, which chanced to have their collecting instinct strongest developed, would make a somewhat larger pile [of rubbish in which to plant its eggs], and the eggs... would in the long run be more freely hatched and produce [more] young (144–5).

"Chance" also appears to govern the question of whether once-useful organs that no longer contribute to the survival of organisms are preserved or disappear. This is the flip side of the appearance of new variations. If a once-useful organ that is no longer useful disappears, that too may be regarded as a variation, and that sort of variation is also due to "chance":

> In considering the eye of a quadruped, for instance, though we may look at the eye of a molluscous animal or of an insect, as a proof how simple an organ will serve some of the ends of vision; and at the eye of a fish as a nearer guide of the manner of simplification; we must remember that it is a mere chance (assuming for a moment the truth of our theory) if any existing organic being has preserved any one organ, in exactly the same condition, as it existed in the ancient species at remote geological periods (151).

In short, wherever Darwin had occasion to go behind the various "apparent causes" of variation to explain exactly how those causes operate to effect change in organisms, he reverted, as he had in the "Sketch," to the notion of "chance." At the same time, he frequently turned to the idea of "unknown causes" or "unknown agencies." These two explanations coexisted in an uneasy tension throughout the "Essay." They lingered on into the early editions of the *Origin*. It appears that by the time of the sixth edition of the *Origin* Darwin was no longer willing to tolerate the ambiguity. As we have seen, he resolved it by essentially dismissing the *word* "chance" from the domain of his theory without dismissing the idea.

CHANCE IN THE *ORIGIN*

Darwin continued to refine his theory in the years after the *Notebooks*, the 1842 "Sketch," and the 1844 "Essay." Between 1844 and 1856 he spent the main part of his scientific endeavors on his pioneering work on barnacles. Chance variation continues to play a role, if a subdued one, in this work. In 1856 he resolved to take up with undivided attention his work on "the species question," culminating in 1859 in the publication of *Origin*.

A review of Darwin's writings from 1839 to 1859 makes two things clear. He did not change his opinion about the role of chance in variation, and he did decide he needed to change his mode of exposition. After the appearance of the first edition of the *Origin* Darwin struggled properly to present his opinion to the public audience. He tinkered, fiddled with, and changed the way he expressed himself. It is necessary, to bring home the full force of these changes, to examine them in some detail. This may seem a tedious exercise, but much is at stake in the question of Darwin's evolving views on this subject, and the evidence is quite revealing.[8] We begin with a passage that remained unchanged through every edition of the *Origin* that seems to assert that "chance" plays *no* role in variation:

> I have hitherto sometimes spoken as if the variations—so common and multiform with organic beings under domestication, and in a lesser degree with those under nature—were due to chance. This, of course, is a wholly incorrect expression, but it serves to acknowledge plainly our ignorance of the cause of each particular variation (*Origin* 275: V. 4–5).

This passage is curiously misleading, in two ways. It asserts, first, that Darwin has "sometimes" spoken of variations as "due to chance," but that is not quite true. In fact, he had hitherto rarely spoken of variation as due to chance (or "accident"), and that only in earlier editions of the *Origin*. By the fifth edition he completely eliminated almost all of these references, leaving only *one* mention of "chance" as the source of variation in his final treatment of the subject prior to his assertion quoted above, and this one he qualified so as to suggest he did not really intend it anyway. I reproduce here some samples of his shifting usage, showing first the matter as presented in the first edition of the *Origin* followed by changes to subsequent editions:

> I. [First through fourth editions]: Bearing such facts in mind, I can see no reason to doubt that an *accidental* deviation in the size and form of the body, or in the curvature and length of the proboscis, &c... [gives] the individual so characterized... a better chance of living and leaving descendants (*Origin* 183: IV. 117, emphasis supplied).
>
> II. [Fifth and sixth editions]: Bearing such facts in mind, I can see no reason to doubt that under certain circumstances individual differences in the curvature or length of the proboscis, &c.... so that certain individuals would be able to obtain their food more quickly than others [and so to survive] (*Origin* 183: IV. *117+8e*).
>
> III. [First through fourth editions]: In such case [where places in the economy of nature are still unoccupied], every slight modification, which in the course of ages, chanced to arise, and which in any way favoured the individuals of any of the species, by better adapting them to their altered conditions,

would tend to be preserved; and natural selection would thus have free scope for the work of improvement (*Origin* 166: IV. 21).

IV. [Fifth and sixth editions]: In such cases [as above], slight modifications, which in any way favored the individuals of any species, by better adapting them to their altered conditions [the remainder of the passage is identical to the original version] (*Origin* 166: IV. *21e*).

V. [All editions]: Mere chance, as we may call it, might cause one variety to differ in some character from its parents, and the offspring of this variety again to differ from its parent in the very same character and in greater degree; but this alone would never account for so habitual and large an amount of difference as that between [well-marked] varieties of the same species and species of the same genus (*Origin* 205: IV. 254; bracketed words added to third and subsequent editions).

For whatever reason, Darwin decided in the first two of these passages (I and II above) to back away in the fifth edition from any indication that variation is due to chance. In the revised fifth edition, he replaced "accidental deviation" with "individual differences," and "every slight modification that chanced to arise" by the more neutral "slight modification which favored the individuals." The latter versions make no mention of chance, or for that matter any cause of variation at all. Some modifications simply "existed" or "favored the individuals," however they may have come about. One must regard these changes as deliberate, reflecting a wish on Darwin's part to avoid the suggestion that variations are somehow "by chance."

The third, unaltered passage, although identifying chance as the source of variation, qualifies that idea significantly. First, Darwin noted that chance is "as we may call it," suggesting that the word is a cover for something else (identified as "our ignorance" in Chapter Five). He thus withdrew "chance" as a candidate for the source of change even as he allowed it. Second, he reduced the role of this "whatever it is" to a minor one by saying that "it alone would never account" for the large differences among species and genera in nature. In other words, something else besides this "whatever it is" must also be playing a role, and a much greater one at that. This "something else" turns out to be natural selection, which here assumes the role of an agent or cause of change alongside other causes of which we are ignorant. The passage shows Darwin was even in the third, most explicit passage about "chance" as source of change most reluctant to go very far with this expression.

The second curious feature about the statement that Darwin has "hitherto spoken" of variations as being due to chance concerns the word "hitherto." In fact, assertions about "chance" or some equivalent expression are more common *after* this passage than before it. But here again we find interesting changes from the first to the sixth edition. I find several mentions of the word "chance" or "accident" (or an equivalent expression) used as a cause of variation later in the *Origin*, and again

Darwin saw in subsequent editions the need to change what he had written in the first edition. I here reproduce a selection of these passages, the first edition's version followed immediately in each case by the transformed version in a later edition:

I. [First through fourth editions]: To sum up... variation is a very slow process, and natural selection can do nothing until favourable variations chance to occur, and until a place in the natural polity of the country can be better filled by some modification of some one or more of its inhabitants (*Origin* 327: VI. 54).

II. [Fifth and sixth editions]: To sum up... variation is a slow process, and natural selection can do nothing until favourable individual differences or variations occur, and until a place in the natural polity of the country can be better filled by some modification of some one or more of its inhabitants (*Origin* 327: VI. *54e*).

III. [First through fourth editions]: But I believe the effects of habit are of quite subordinate importance to the effects of the natural selection of what may be called accidental variations of instincts;—that is of variations produced by the same unknown causes which produce slight deviations in bodily structure (*Origin* 382: VII. 28).

IV. [Fifth and sixth editions]: But I believe that the effects of habit are in many cases of subordinate importance to the effects of the natural selection of what may be called spontaneous variations of instincts;—that is of variations produced by the same unknown causes which produc slight deviations of bodily structure (*Origin* 382: VII. *28e*).

V. [Third edition]: Why has not the ostrich acquired the power of flight? But granting that these organs have happened to vary in the right direction, granting that there has been sufficient time... who will pretend he knows the natural history of any one organic being sufficiently well to say whether any particular change would be to its advantage (*Origin* 229: IV. *VII. 382.46c*).

VI. [Fourth through sixth editions]: Why has not the ostrich acquired the power of flight? But granting that these organs have varied in the right direction, granting that there has been sufficient time [the rest of the passage unchanged] (*Origin* 229: IV. *VII. 382.47e*).

VII. [First edition]: Hence, we may conclude, that domestic instincts have been acquired and natural instincts have been lost partly by habit, and partly by man selecting and accumulating during successive generations, peculiar mental habits and actions, which at first appeared and from what we must in our ignorance call an accident (*Origin* 389: VII. 80).

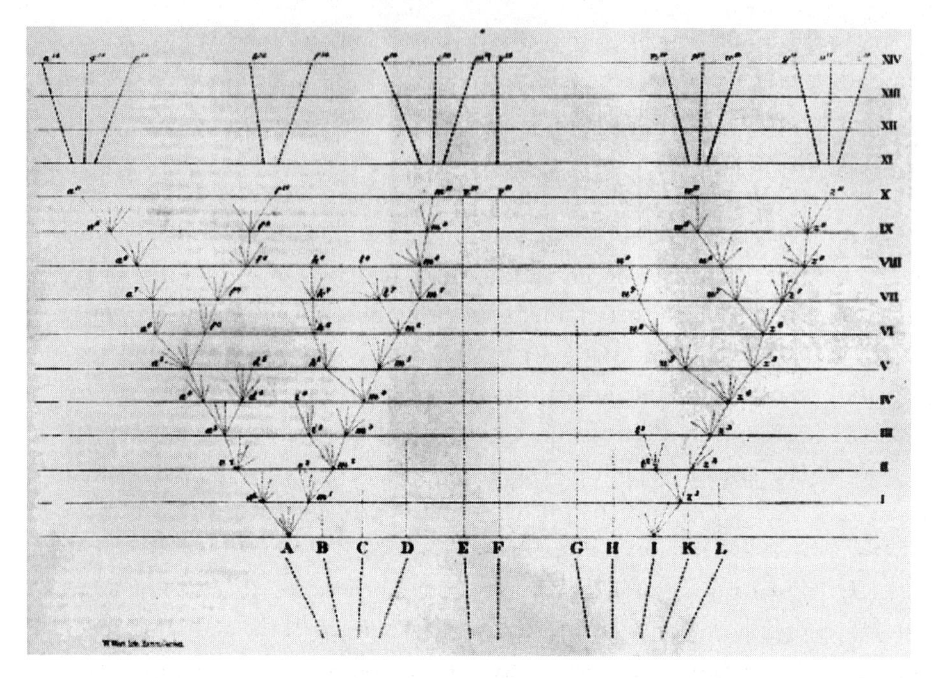

FIGURE 5.1 Darwin's most famous visual representation of the evolution of life's diversity. This is the only diagram to appear in the *Origin of Species*. He retained it unaltered through all six editions.

VIII. [Unchanged in subsequent editions]

In these passages we find in later editions again the same drawing away from "chance" or "accident" or "happenstance" that we saw in earlier editions. In the first three instances (I–VI) the fifth and sixth editions remove the word and idea of chance (accident) altogether. In the fourth (VII–VIII), while "accident" is retained, it is so qualified as simply another word for "our ignorance" as to render it inoperable.

In addition we here find, in the second passage above (paragraph IV), the introduction of a new word that replaces chance: "spontaneous variation." The phrase does not appear in any edition of the *Origin* prior to the fourth. Here it appears only rarely (e.g., *Origin* 443, 445: VIII. *159.1: d–f; 159.13: d–f*). By the sixth edition it has become a prominent idea, mainly in Chapter VII, "Miscellaneous Objections to the Theory of Natural Selection," the only new chapter Darwin added to his original text. In this chapter Darwin wished to address the criticisms his book had received from prominent naturalists in the years since its first publication in 1859. His chief concern was with the young biologist St. George Mivart, who had published a spirited attack against the *Origin* in several articles in the Catholic periodical *Month* in 1869.[9] Darwin saw a need to address these attacks and so added his Chapter VII, which became a permanent part of the *Origin*.

THE *ORIGIN* AND THE AFTERMATH

A decade after the first publication of the *Origin* in 1859 the firestorm of criticism only intensified, thanks mainly to the critical reviews by St. George Mivart that started to appear late in the 1860s. Mivart made Darwin's "fortuity" a centerpiece of his criticisms of the theory.[10] These reviews prompted Darwin to make further adjustments to his presentation of "chance" in later editions of the *Origin*, as we shall show. Most of the changes center on decisions he was making about how to disguise or render innocuous the role of "chance" without really removing it from its central position in his theory.[11] For example, a builder ("architect" in Darwin's usage) makes use of stone fragments that have fallen from a precipice to fashion a beautiful new building. But the stones fell into their various shapes and sizes not to suit the builder; rather, they suited the builder because they chanced to have the shapes and sizes they did.

In the same way, variations in organic beings arise not with the end of survival in view. Rather, variations just happen, and then either serve survival needs or not depending upon what variations happened to occur in particular natural environments. Processes like these are unpredictable—who can possibly know what shapes and sizes the stone fragments will assume when they fall from a precipice; who can possibly predict what variations nature will throw up among her vast production of new organisms, and when, and where; the variations are blind—they do not "look ahead" to see what will be needed for future use, but will be used or not according to what they happen to be. Causes for such outcomes exist, without doubt. But one is still justified in saying such outcomes are "by chance" because of the near if not total impossibility of saying what they will be before they actually occur.

Mivart believed Darwin vastly overstated the case for natural selection in effecting change, and also that Darwin's theory failed to account for many natural phenomena. These criticisms forced Darwin to return to the issue of "causes of variation," as he perhaps realized that was the most vulnerable part of his theory, insofar as he had yet failed to give a completely consistent or coherent account. Mivart was especially critical of Darwin's reliance on the idea of "fortuity" (as Mivart called it) in causing variation. Darwin had already been under some pressure from several friends to admit that variation must be guided by a higher power or creative law. Now Mivart was making a public spectacle out of Darwin's "fortuity." Darwin had to fight back. He did so not by defending the role of chance in variation but by deleting most mentions of it. By the fifth edition, "chance" and "accident" have almost entirely disappeared from the *Origin*.

Two new expressions now took the place formerly occupied by chance. The first is the expression "laws of growth," implying that chance could play *no* role in variation *(Origin* 234: IV. *VII. 382. 65. 0. 12. 5:f*; ibid., 237: IV: *VII 382. 65. 0. 22. 1:f*; ibid.,

239: IV. *VII. 382. 65. 0. 40. 1:f*; ibid., 243: IV. *VII. 382. 65. 0. 50. 19:f*). If "laws" govern all change where is the room for chance? At the same time, these "laws" are still only "dimly seen" (*Origin* 234: IV. *VII. 382. 65. 0. 12. 1–3:f*), or even more strongly (and frequently), that they are "unknown" (*Origin* 234: IV. *VII. 382. 65. 0. 12. 7:f*)), and that we are "quite ignorant" of what they are (*Origin* 238, 240: IV. *VII. 382. 65. 0. 33. 2:f; 382. 65. 0. 43:f*).

The other is the phrase "spontaneous variation." This is new in the fourth edition, and raises the question of what Darwin may have intended by it. I will offer here only three brief observations, postponing a fuller discussion of this subject to a later chapter. The first is that "spontaneous variation" is sometimes referred to as "so-called spontaneous variation" (*Origin* 232: IV. *VII. 382. 61. 0. 0. 5:f;* ibid., 234: IV. *VII. 382. 65. 0. 12. 6:f*), again suggesting, as was true with "chance," that the phrase is a mere expression, standing in for something different (though unknown). Secondly, Darwin contrasted "spontaneous variation" with variation that comes about in circumstances in which "the nature of [external] conditions" plays a decidedly more important role (*Origin* 234: IV. *VII. 382. 65. 0. 5: f*). In other words, spontaneous variation is Darwin's answer to Lamarck: in some cases external conditions cause variation, as Lamarck argued, but at other times variation "just happens," regardless of conditions. They happen "spontaneously," a word in this context that suggests "chance" without actually saying so. Finally, it is worth noting that "spontaneous variation" is listed alongside other causes of variation—correlations of growth, use and disuse, and direct and indirect environmental influences—as just one more of the ill-understood "laws of variation." Whose place has it taken from earlier accounts? Apparently, the place of "chance."

Where did "spontaneous variation" come from? Darwin does not say, so any answer must be in some measure conjectural. I attempt some educated guesses in chapter 6. As a preliminary observation, though, Darwin's new phrase "spontaneous variation" can be seen as part of the larger strategy in later editions of the *Origin* to banish the *words* "chance" and "accident" as much as possible from an active role in his theory. Where "chance" and "accident" once stood, now "spontaneous variation" stands. Darwin had obviously discovered a phrase with which he was more comfortable. Did the substitution signal a shift in his basic outlook about variation?

The answer is, probably not, and certainly not when chance is understood as "blind with respect to future adaptive needs." In 1861 Darwin confessed to a correspondent that chance was more or less the inevitable consequence of his disbelief in both design and Lamarckism:

> My greatest trouble is not being able to weigh the direct effects of the long continued action of changed conditions of life without any selection, with the

> action of selection on mere accidental (so to speak) variability.—I oscillate much on this head but generally return to my belief that the direct action of the conditions of life has not been great. At least this direct action can have played an extremely small part in producing all the numberless and beautiful adaptations in every living creature (*CCD* vol. 9, 107, letter to T.W. St C. Davidson, April 1861).

The implication here is that if variations are not by "direct action of conditions" they must be by "so-to-speak accident" (i.e., chance). Therefore they cannot have been produced by "designed" laws, final causes, or anything else. "Chance causes" would not mean just "we do not know," but rather that "we *cannot* know, in principle, and not just because the divine cause is imponderable, but rather because no Designer exists as a higher cause behind the actual forces at work in nature."

NOTES

1. Cf. Kohn and Hodge (Kohn, 1985), Chapter 6, for a discussion of chance variation in the early *Notebooks*. I argue that they are right to defend a Darwinian turn to chance variation in the *Notebooks*, and are right in associating it with Darwin's pivotal discovery in 1839 of the analogy between selection under domestication and selection under nature. I place the former discovery, chance variation, earlier than they do (i.e., in 1837 rather than 1839).

2. The Aristotle passage quoted in the "Historical Sketch" raises the question whether the teeth of cows were made "for chewing," or were made by accident and then turned out to be useful for their survival. Darwin attributes the explanation from "accident" to Aristotle himself, whereas in fact Aristotle was reporting the views of Empedocles, whose views he then proceeded to dismiss as incorrect. Aristotle maintained the teleological perspective. This issue is discussed in Johnson (2007). Cf. Gotthelf (1999); Stott (2012, Chapter One); Lennox, 1993.

3. Ospovat (1981) documents Darwin's shift from a "perfectionist" view of adaptation to a "what happens to work best" view. The insight occurred to Darwin in 1838, earlier than Ospovat's placement of it.

4. The "Sketch" and the "Essay" appear in a modern volume, *Evolution by Natural Selection*, edited by Gavin de Beer (1958).

5. The effect of conditions on reproductive organs needs to be distinguished from change caused somehow by the "act of reproduction," a cause that figures with some prominence in the "Sketch" but that, as Francis Darwin pointed out, is not admitted as a separate cause in the *Origin* (53 n. 1). In any case, variation resulting from the "act of reproduction," even in the "Sketch" is "vastly increased when parents [are] exposed for some generations to new conditions" (53). The meaning seems to be that "new conditions" over time cause changes in the reproductive *organs*, Darwin's fundamental position in the *Origin*.

6. This reading seems to be the one adopted by Francis Darwin in his notes to the "Sketch." He expresses uncertainty about why Darwin ruled out "chance" in his reference to Lamarck, noting that Darwin had already argued that "variation is ultimately referable to conditions and that the nature of the connection is unknown, i.e., that the result is fortuitous" (49 n. 1).

7. Cf., 97, 107, 114–15, 119, 130, 150, 226–8, 241. Cf. Schweber's (1985, 56–7, in Kohn, 1985) illuminating discussion of "chance variation" in the 1844 "Essay."

8. One must disagree with James Lennox (2004), who wrote: "Darwin, in fact never refers to 'chance variations' in the *Origin*, though occasionally he will note that if a beneficial variation 'chances [i.e. happens] to appear', it will be favored by selection (see 37, 82)." Perhaps Lennox means only that Darwin did not use the *exact expression* "chance variations" in the *Origin*, and I think he is right about that. But it is evident that Darwin in fact frequently made the point that chance variations occur in nature, or at least that he did so in the first edition of *Origin*. For later editions he worked hard and quietly to expunge the expression while retaining the idea (Cf. Johnson, 2010, 1–27).

9. Several of Mivart's reviews have been collected into a single volume *On the Genesis of Species* (1871).

10. This is a simplification. While Mivart did question "fortuity" in the theory, he also faulted Darwin for other sins: giving natural selection too large a role in evolution (and Mivart was an evolutionist), being "too dogmatic" in the presentation of his views, and presenting a theory that ruled out a role for an omniscient designer. Darwin started to get the scent of these criticisms only after he read Mivart's 1871 work *On the Genesis of Species*. Darwin and Mivart exchanged letters on these matters in early 1871 (*CCD*, vol. 19, 30–7). Darwin's answers to Mivart's criticisms were: (1) that he had *always* bent over backward to allow a due role for Lamarckian "use/disuse" in his theory (for more on this, see chapter 8 of the present work; and *CCD* vol. 19 31–2 and nn. 3 and 7); and (2) that if he had been "dogmatic" in his published works he would be grateful to Mivart to say more clearly where he had made this error—for certainly, Darwin thought, dogmatism in science *is* an error. Mivart made token adjustments in subsequent writings to address these two concerns, but looking over how he changed his expression one must conclude that his revisions were half-hearted and perhaps insincere. His statements reveal his real concern was with the "atheism" that Darwin's theory seemed to entail.

11. One observes in the *Notebooks* that Darwin was open to disguising his true opinions about questions bearing on his religious views to a larger public. Not only did he occasionally remind himself to be cautious in how he expressed himself, he also on at least one occasion told himself to conceal his self-avowed "materialism": "To avoid stating how far, I believe, in Materialism, say only that emotions, instincts degrees of talent, which are heredetary [*sic*] are so because brain of child resembles parent stock.—(& phrenologists state that brain alters)" (*CDN* M-57).

6

"So-Called Spontaneous Variation"

SOMETIME BETWEEN 1861 and 1866 Darwin decided to make a "silent" change to the *Origin*, one that at first may appear innocuous but on closer inspection seems significant: for the words "chance variation" (or sometimes "accidental variation") he substituted the expression "so-called spontaneous variation." The latter makes its first inconspicuous appearance in the 1866 fourth edition (three times), then is brought more actively into play in the fifth edition (1869), and finally assumes almost the entire weight of what had previously been "chance variation" in the sixth (1872), the last edition Darwin edited. By the sixth edition the expressions "chance variation" and "accidental variation," formerly common, had all but disappeared.

Why Darwin made these alterations to the *Origin* can only be guessed, because he did not say. The chances (so to speak) are good, though, that he was, as often in later editions of the *Origin*, reacting to criticisms from his readers. One such criticism was a worry that "chance" could be any part of the evolutionary process. By this time (1868), and actually much earlier, as we have seen, he believed it was. The question had become: how to present this idea in a less worrisome way? The phrase "so-called spontaneous variation" was part of Darwin's answer.

As we have seen, "chance" was for Darwin a correct way to characterize not just some but most variations, in the sense that variations, he thought, were usually random with respect to the needs of organisms that were born with them (the Geoffroyian and Lamarckian causes of "direct action" of conditions on organisms and use-inheritance, which Darwin did accept as causes of some variations, are exceptions). Variations had to be "in the right direction at the right time" for them to be selected, but no guarantee exists that this will always or even usually happen. In fact the odds run the other way: most variations will be either neutral in fitness value, and so will be "ignored" by natural selection, or positively harmful, in which case natural selection can be counted on to destroy them (*CCD* vol. 8, 340 [September 1, 1860], to C. Lyell and 355 [September 12, 1860], to C. Lyell).

The problem is, many of Darwin's readers did not grasp this distinction, and even if they did it is not clear that they would have cared. Many critics, even friendly ones like Asa Gray and Charles Lyell, wanted a role for designed directionality in variation, not "blindness" with respect to future adaptive needs. Darwin never could accept this. If his theory required anything like a creational law or force, he wrote to Lyell, he would reject it as "rubbish" (*CCD* vol. 7, 345 [October 11, 1859], to C. Lyell). For Darwin, too many variations are non-adaptive in fitness value or even harmful, leading to the death of organisms. Others have no obvious value for the survival of the organism but serve only to give scope for human fancy and the breeder's art. Why would an intelligent designer do these things, Darwin wondered? Nevertheless, he did not want to be understood as claiming that the organic world is without rhyme or reason, and he saw that his reliance on the words "chance" and "accident" were causing people to read him that way. This seems to explain the substitution of "so-called spontaneous variation" for the more controversial words.

"So-called spontaneous variation" does have advantages over the words "chance" and "accident" if one is trying to soothe the qualms of anxious readers. For one thing, it seems less offensive to theistic conceptions, in the sense that variations can be "spontaneous" and nevertheless designed; "chance" or "accident" makes that possibility more difficult to imagine. Also, "spontaneous variation" would probably be regarded as more compatible with "law-governed" or "caused" than the other expressions while still not suggesting what Darwin could not suggest, what those laws or causes are. Finally, by adding the qualifier "so-called" Darwin was again tacitly defusing the criticism that "chance" was still part of his theory by making it seem like he was only saying that "spontaneous variation," while perhaps suggesting "chance," is nevertheless "incorrect," but is a suitable substitute for "our ignorance."[1]

DARWIN'S SOURCES

Our question is, how did Darwin come up with that particular expression? It is possible that he invented it himself. That explanation, though, seems unlikely in view of the fact that the phrase was in circulation already, in fact in several authors with whose writings Darwin was familiar. We will examine these other possibilities and then venture an opinion about the most likely source.

The question is complicated by another one: what did his source (if he had one) *mean* by the expression? Did Darwin merely adopt a phrase whose established meaning was already congenial to him, or did he put his own "spin" on it? The examination thus needs especially to take a look at issues both of meaning and timing—when a particular source appeared, what the phrase meant to its author, when Darwin read it, and when he began employing the new expression in his own writing.

It is to be emphasized that our question does not concern the origin of the *idea* of "chance" or "so-called spontaneous" variation. That, as shown in previous chapters, is a more complicated question. Let it suffice here to recall only that the idea was deeply engrained in biological thought well before Darwin. He may have picked it up from any number of sources from Aristotle[2] to Francis Bacon to David Hume to Maupertius; or, for that matter, he may have come upon it on his own. Whatever its source, the idea already had found a place in Darwin's theory as early as 1837, as is confirmed by several entries in the *Notebooks* and as documented above.[3]

As for the *expression* "spontaneous variation," in contrast to the idea, three candidates stand out as the most likely sources: Prosper Lucas, Herbert Spencer, and Aristotle. I shall examine each in turn directly. But as a preliminary matter, four other people who could plausibly be assumed to have influenced Darwin should be ruled out: Darwin's friend and confidant J. D. Hooker, his mentor Charles Lyell, his "bulldog" T. H. Huxley (also called Darwin's "general agent" by Huxley's son Leonard Huxley in his *Life and Letters of Thomas H. Huxley*, 1901, vol. 1, 183), and the increasingly well-established philosopher J. S. Mill.

Of the four, only T. H. Huxley and Mill appear ever to have used the expression. Huxley did so in a letter to J. D. Hooker, written on April 18, 1861. In it he mildly criticized Darwin for having to speak "as if" variations are "spontaneous or a matter of chance" because Darwin could not make out the "law of variation" that would explain how variations arise (1901, vol. 1, 245). At the same time Huxley understood that Darwin's formulation—chance variation—is what mainly separated Darwin's theory from that of Lamarck, and Huxley clearly understood Darwin's way of putting it to be an improvement over Lamarck. Darwin had not used the phrase "spontaneous variation" before this letter was written, so one should weigh the possibility that he picked up the phrase from Huxley, or from Huxley's correspondent Hooker.

No evidence can be found to support that speculation. Huxley's letter was sent to Hooker, not Darwin, and it is impossible to say that Darwin ever saw it. Nor do we have any letters or other correspondence from Hooker to Darwin in which he may have passed along the phrase. In fact, Hooker steered clear for the most part of engaging Darwin in the big question here—the cause(s) of variation. The closest he came to discussing this issue with Darwin was to point out that some of Darwin's readers, notably Charles Lyell, had confused "natural selection" and "variation." Darwin responded to this confusion by developing his architect metaphor, to be discussed in the following chapter. But as he worked through the metaphor, Darwin never did employ the expression "spontaneous variation." Invariably, he preferred a term of much older vintage in his vocabulary: accident. Darwin's three closest friends thus are not plausible sources for "spontaneous variation."

J. S. Mill, in his *Utilitarianism* (2nd Edition, 1864), comes into the discussion of possible influences on Darwin because of the following passage in that work:

> The moral faculty if not part of our nature is a natural outgrowth from it capable like other acquired faculties in a certain small degree of springing up spontaneously (quoted in a letter from Darwin's son W. E. Darwin to Darwin in 1871 [*CCD* vol. 19, 244]).

The idea expressed by Mill, referring to the spontaneous origin of the "moral faculty," is congenial to Darwin's expression "so-called spontaneous variation" that he introduced into the *Origin* beginning in 1866 to describe how some structural variations come about. But we must rule out Mill as a possible source. Darwin left no evidence in his *Reading Notebooks* that he ever read Mill's *Utilitarianism*, certainly not by 1864 or earlier. No copies of Mill's works are to be found in Darwin's library at Down or at the Cambridge University Library. It is more likely that Mill picked up the idea from Spencer or someone else than that he invented it, and so more likely that Darwin similarly first found the phrase in another writer than in Mill. Even if Mill devised the expression on his own, his use would not have been noticed by Darwin.[4]

PROSPER LUCAS

The first place that "spontaneous variation" shows up in the literature that Darwin had read prior to the *Origin*, as far as I can tell, is Prosper Lucas's 1847–1850 work

Traite Philosophique et Physiologique de L'Heridite Naturelle (2 volumes, 1847–1850). Lucas was a French physician and medical writer whose research was centered on questions about heredity, especially the hereditary transmission of mental illnesses. In the course of his work he devoted some attention to what he took to be the causes of variation. Darwin read his book between September 5 and November 15, 1856 (*CCD* vol. 4, 495 [128:20] in *Reading Notebooks*) and made extensive marginal notations (*Marginalia* 513–23), including notations on passages in which Lucas discussed variation and heredity. That Darwin retained interest in and memory of Lucas's writings is confirmed by his employment of many of Lucas's examples in his own 1868 work *Variation of Animals and Plants under Domestication* and again in the 1871 work *Descent of Man*.[5] Since Darwin started compiling his notes for the former book in 1860 and began writing it in 1862, it is not hard to believe that Lucas's employment of the phrase "modification... sans cause [et] spontanée" may have caught Darwin's eye.[6]

Lucas was interested in knowing, among other things, whether acquired characters are inherited. Some earlier writers, he noted, thought that they are not (e.g., J. C. Prichard in his *Researches into the Physical History of Man*),[7] although many others believed that they are (notably E. Geoffroy St. Hilaire and Lamarck).[8] Lucas himself singled out two primary causes of variability in organisms. One such cause he called "immediate," by which he meant "induced by external conditions." (External conditions, in turn, are of two kinds—"physical" and "moral.") The other he called "mediated" or "congeniale" (i.e., congenital). These, in contrast to the first, arise "independently of circumstances," in such a way that we are permitted to say they come about "in a spontaneous manner" (*Traite*, vol. 2, 442–5). Darwin heavily marked these passages, and translated Lucas's "modifications spontanées" by the words "spontaneous variations" in his marginal annotations to Lucas's work (*Marginalia* 521).

Lucas's theory was not particularly original, nor did he really claim that it was. He acknowledged the ideas of Lamarck and Geoffroy when describing the action of "external circumstances" in inducing modifications, and accepted their ideas of the heritability of acquired characters, that is, characters acquired after birth. Darwin took him to be saying that some variation is "the direct effect of external agencies on the parents and on the individuals after they have life, or when born" (*Marginalia* 521). Nor did Lucas seem to think his idea of "spontaneous variation" was new with him either. It is, in fact, an echo of Lamarck's "power of life," what Darwin came to call Lamarck's "progressivism," the supposed innate tendency of organisms to ascend the ladder of progressive complexity of organization. Lucas himself came to call this "innate" principle of change "*l'unique impulsion d'une des lois de la vie*" (*Traite* vol. 2, 445). This expression shows that

Lucas did not mean by "spontaneous" the very different idea of "uncaused," but only the more limited notion that some variations are not directly induced by external conditions.

What is new with Lucas, though, is the *expression* "spontaneous modification." This is not to be found in the writers Lucas cited in his discussion of variation, although Lamarck (and many others, as Lamarck noted) had employed the expression "spontaneous *generation*" to account for the origin of life itself (cf. *ZP*, 236–8).[9] Spontaneous variation is, however, quite different from spontaneous generation, and Lamarck had not used the former expression. On the contrary, Lamarck (and Geoffroy and Prichard and the other authors mentioned by Lucas) were quite sure variations typically arose in organisms *after* they had been born, and were caused either by the direct effects of external conditions on organization, or, in the case of Lamarck especially, by the operation of an innate "power of life" that directed organisms upward on a scale of increasing complexity. Neither of these causes could properly be called "spontaneous" in Darwin's sense, blind with respect to future adaptive needs, and especially in the sense, never accepted by Darwin either, of "uncaused." For all authors prior to Lucas variations from whatever source were both "caused" and of positive adaptive value for organisms acquiring them.

HERBERT SPENCER

Herbert Spencer (1820–1903) comes into the discussion of spontaneous variation because of a letter he wrote to Darwin on February 22, 1860, just two months after the first edition of the *Origin* had appeared. In it Spencer acknowledged that Darwin had "wrought a considerable modific[ation] in the views I had held." Specifically, Darwin had caused Spencer to see that "direct adaptation [to external conditions]" could not explain all organic evolution (*CCD* vol. 8, 98 [February 22, 1860]), and that some other mechanism must be involved. At first one might think that the new mechanism introduced by Darwin is "natural selection." But in fact Spencer claimed to have already described the operation of this, what he later came to call "survival of the fittest," at least in human populations, in his 1852 publication "A Theory of Population" in the *Westminster Review*:[10]

> Many (&c) must have been struck [before the *Origin*] with the fact that among all races of organisms the tendency was for the best individ[uals] only to survive & that so the goodness of the race was preserved. I have in Essays on Population &c remarked this as a cause of improvement among mankind (*CCD* vol. 8, 98 [February 22, 1860]).

While he thus was tacitly claiming priority for the insight about the operation of natural selection,[11] Spencer had not been able to explain how the "more fit" and the "less fit" arise in the first place:

> But I & every one overlooked the selection of "spontaneous" variations without which I think you have clearly shown that many of the phenomena are insoluble (*CCD* vol. 8, 99 [February 22, 1860]).

Darwin did adopt Spencer's expression "survival of the fittest," adding it to the *Origin* beginning with the fifth edition (1869) as a synonym for natural selection.[12] It may, then, have been Spencer from whom Darwin took the expression "spontaneous variation." If that is so, what is the evidence? But first one may ask where Spencer got the phrase. The fact that he placed the word inside quotation marks suggests he may have borrowed it himself, but if he did it could not have been from Darwin, for Darwin had not yet employed the expression by February 1860.

One possibility is that Spencer was repeating something he had previously written himself. In addition to his "Theory of Population," Spencer had published, also in 1852, his "Development Hypothesis" (*The Leader*, March 20, 1852). In this work he outlined his case for gradualist transmutation of species, or "evolution," contrasting his theory with all of those who still argued for separate and special individual creations. Darwin encountered this short essay in a collection of Spencer's essays[13] sometime soon after the appearance of the first volume of that work in 1858. He wrote to Spencer on November 25, 185, to thank him for "the kind present," and mentioned specifically his admiration for the "so-called Development Theory" (*CCD* vol. 7, 210). He retained or recalled memory of Spencer's theory after the *Origin* first appeared, for on February 2, 1860, he wrote to Spencer asking for a precise citation to the essay as it first appeared in the *Leader* (*CCD* vol. 8, 66 [February 2, 1860]). Darwin wished to include a paragraph on it in the "Historical Sketch" that he prefaced to the third edition of the *Origin*, indicating that Darwin saw Spencer as a forerunner in some respects to his own theory.[14] But "spontaneous variation" does not appear in the "Development Hypothesis" or in the "Theory of Population."[15] One gets the impression from Spencer's February 22, 1860, letter to Darwin that Spencer regarded Darwin as the discoverer of the idea of "spontaneous variation," even if Darwin had not called it by that name, rather than that Spencer was referring to his own prior theory or to anything he had previously written.

Spencer's theory of development as it existed in 1858 was in part a polemic against special creationists and in part a defense of Lamarck's "transmutationist" theory, or at least that part of it known as "use-inheritance" (although Spencer did not mention

Lamarck by name). This is the theory that attributes modification of organisms to the "direct action" of changing external conditions, organic and inorganic, on "habits" (or "use/disuse" of parts); these in turn giving rise to new structures by small and incremental steps. Spencer encapsulated the essence of his theory in the same February 22, 1860, letter to Darwin in which he first employed the phrase "spontaneous variation":

> I was under the erroneous impression that the *sole* cause [of modifications in organisms] was adaptation to changing conditions of existence brought about by [new] habit, using the phrase conditions of existence in its widest sense as including climate, food, & contact with other organisms [Spencer here cites his earlier essay on the "Development Hypothesis"]. You have convinced me that throughout a great proportion of cases, direct adaptation does not explain the facts, but that they are explained only by adaptation through natural selection (*CCD* vol. 8, 98, emphasis supplied).

The last sentence implies that Spencer believed the novelty in Darwin's theory to be "natural selection," but we have earlier seen that Spencer himself (and, he thought, many others) had already observed the "tendency of the best [i.e., most fit] individuals only to survive." Darwin's novelty, rather, was his "explanation" of the origin of many variations as being caused by what is sometimes called "chance" (i.e., spontaneity). There is no room in Spencer's "development hypothesis" as it appeared in 1858 for "chance" variation. Variations are to be understood as "direct" adaptations to changing conditions. Apparently, then, when Spencer credited Darwin with having been the first to point out the importance of "spontaneous variations" in the modification of species in his February 22, 1860, letter, he was simply using a new expression to render what up until then had been called by Darwin "so-called chance variation."

Nothing else in Spencer's writings prior to February 1860 provides evidence that Spencer borrowed the phrase "spontaneous variation" from someone else. If we then assume that Spencer invented the expression, we may next ask whether Darwin borrowed it from him. He could not have taken it from any writing of Spencer's that was published prior to 1860, since that is evidently the first time Spencer used it—in his letter to Darwin. It is possible, though, that Darwin did not recall the expression from Spencer's letter, but rather borrowed it from a *later* work by Spencer, one that appeared after the *Origin* was first published. Recall that Darwin's first employment of the phrase as a substitution for "chance variation" came in the fourth edition of the *Origin*, published in 1866. Thus we must countenance the possibility that Darwin was drawing on some source by Spencer that appeared some time *after* the

1860 letter but *prior* to his revisions to the fourth edition, which would mean up to December 1866.

The only candidate from Spencer's *oeuvre* that fits the bill is his *Principles of Biology* (1864–1867), the first volume of which was issued in installments beginning in January 1863 and continuing through the end of 1864 (see *CCD* vol. 12, 397, n. 5 for a brief history of the publication record of Spencer's work up through October 1864). Darwin did read this work as it appeared in successive installments in 1864, as he indicated to Hooker in a letter of November 3, 1864 (*CCD* vol. 12, 391). It is thus worth asking what Spencer had to say about "spontaneous variation" in this work.

The phrase does appear throughout the *Biology*, 11 times, and almost always, interestingly, in quotation marks, just as it had appeared in the February 1860 letter to Darwin.[16] On one occasion, Spencer even used the phrase "so-called spontaneous variation" (*Principles* 167), just as Darwin usually did when he incorporated the expression into the *Origin* in 1866. The suggestion becomes strong, then, that Darwin took the expression directly from Spencer's work on biology.

But a closer look casts doubt on that supposition, for two reasons. The first is that Darwin's *first* use of the expression appeared long before he read Spencer's *Biology*. It was employed by Darwin in a letter to Leonard Horner written in February 1861, one year after Darwin received Spencer's 1860 letter, explicitly to render "accidental variation":

> Man does not cause any variation, he only accumulates any which occur: I do not suppose that God intentionally gave the parent Rock-Pigeon a tendency to vary in size of Crop, so that man by selecting such variations should make a Pouter; so under nature, I believe variations arise, as we must call them, in our ignorance, accidentally or spontaneously, & these are naturally selected or preserved (*CCD* vol. 9, 28 [February 14, 1860]).

This is the only place I have been able to trace that Darwin used the word "spontaneously" to modify "variation" prior to January 22, 1865 (where it appears in a letter to Lyell: *CCD* vol. 13, 35). But the fact that he used it at all and that he explicitly equated it with "accidental variation" to mean "unknown cause" shows that it was established in his mind long before he read Spencer's *Biology*.

The second reason for doubting Spencer's *Biology* as Darwin's source is that Spencer intended the phrase to mean something different from what Darwin intended. Darwin used the expression, as noted above, to stand in for "an unknown cause, a cause about which we are ignorant." Some variations "just happen," and we do not understand why. Spencer agreed with Darwin up to a point. He was as sure as Darwin was that all variations are "caused": "I hold in common with Mr. Darwin that there must be some cause for these apparently-spontaneous variations" (citing Darwin's statement at the beginning of Chapter Five of the *Origin* that "chance [variation] is a wholly incorrect expression [if it is taken to mean 'uncaused']," *Principles* 264). Spencer also, therefore, accepted "spontaneous variation" as an accurate way to render "chance variation" as Darwin employed it in the *Origin* prior to the fourth edition.

But Spencer differed sharply with Darwin about whether the causes of variations that arise by "so-called spontaneity" are really unknown or unknowable.[17] Spencer's theory, unlike Darwin's, gave an explanation for *all* variations—hence his use of the expression "so-called spontaneous variation." The reason for the quotation marks and the qualifier "so-called" in Spencer's formulation is that Spencer did not think variations resulting from what "some people" (i.e., Darwin?) called "spontaneity" were unexplainable: "it seems to me," Spencer wrote, "that [unlike Darwin's belief] a definite cause is assignable [for these apparently spontaneous variations]" (*Principles* 264–5). Spencer went on to spell out where and how he departed from Darwin's theory:

> On the one hand organisms in complete equilibrium with their [external] conditions cannot be changed except by change in [those] conditions…On the other hand, any change in [external] conditions can affect an organism only by changing the actions going on in it—only by altering its functions. The alterations in functions being necessarily towards a re-establishment of the equilibrium…it follows that the structural alterations directly caused, are [functional] adaptations.…We must say that in all cases, adaptive change of functions is the primary and ever-acting cause of that change of structure which constitutes variation; and that the variation which appears to be "spontaneous," is derivative [of] and secondary [to that primary cause] (*Principles* 271–2).

What Spencer had offered, in other words, was an account of what "spontaneous variation" really boils down to, and it is not "unknown cause," as it was for Darwin.[18] His account to this point closely resembles Lamarck's "adaptationism," that is, changes in conditions cause changes in habits (use/disuse or, in Spencer's terminology, function), with these functional adjustments in turn giving rise to new organic structures. This is, surprisingly, nearly identical to Spencer's 1852 theory as set out in the "Development Hypothesis," with some new additions. Spencer in fact now acknowledged Lamarck, as he had not done in his 1852 essay, as having got this much right, faulting him only for failing to explain what caused organisms to "endeavor" to meet the challenge of changing conditions (*Principles* 403–9). Lamarck's failure to give an adequate answer was "question-begging" for Spencer. Spencer's argument is that "spontaneous variations" are "so-called" just because a straightforward explanation *can* be given for them, namely, functional adaptations to changing external conditions; and further, that he (Spencer) understood, as Lamarck, Erasmus Darwin, and Charles Darwin did not, how such functional adaptations take place (involving his theory of cosmic "forces" and the maintenance of cosmic "equilibrium").

The credit Spencer had given Darwin in 1860 for discovering the crucial importance of "spontaneous variation" he now seemed to be taking away. Perhaps Darwin's "chance variations" had prompted Spencer to think about a non-Lamarckian pathway to modification of species, but by 1863–1864 he had become clear that "chance" (or "spontaneity") in Darwin's sense was incorrect. Confirmation of this inference is found in an "Appendix" that Spencer added to the *Principles of Biology* in December 1868, two years after Darwin had first used the expression "so-called spontaneous variation" in the *Origin* but before he had made it the cornerstone of his account of variation. In the "Appendix" Spencer attempted to refute criticisms that his embrace of "evolution" (the derivation of new organic forms from earlier ones) implied his embrace of "spontaneous generation" (the creation of organic from inorganic matter). The two cases are wholly distinct, he argued. The latter could not be explained by any known biological process, and therefore Spencer would not accept it in his science. The expression "spontaneous variation," on the other hand, must be rejected for a different reason: unlike spontaneous generation, spontaneous variations are not really "spontaneous" at all. Their causes can be known, and in fact are known (by Spencer). Spencer even thought Darwin agreed with him about this:

> The very conception of spontaneity is wholly incongruous with the conception of Evolution. For this reason I regard as objectionable Mr. Darwin's phrase "spontaneous variation" (as indeed he does himself); and I have sought to show that there is always an assignable cause of variation (*Principles* 480).

Whether Darwin borrowed the expression "spontaneous variation" from Spencer or not, this comment is strange. Spencer now finds the expression "objectionable," even though he had earlier (in 1860) used the very same expression himself as expressing his views. Apparently Spencer believed that Darwin found the phrase "spontaneous variation" objectionable because Darwin confessed even in the first edition of the *Origin* that "chance variation" is a wholly incorrect expression. Darwin meant that all variations are caused, but that sometimes the cause(s) are not known, and so we say "they arise by chance." Spencer's objection was quite different. "Spontaneous variation" is an incorrect expression because it suggests an unknown cause, whereas the causes are always known, and they are the Lamarckian causes of use/disuse (Spencer's "functional adaptation") and the inheritance of acquired characteristics. In a sense Spencer merely strengthened the Lamarckian account by claiming to explain more fully how use-inheritance works. In so doing he completely transformed what Darwin had meant by his expression "so-called chance variation."

DARWIN'S OPINIONS ABOUT SPENCER

What did Darwin think about Spencer's theory? His marginal notations to the *Biology* reveal that he in fact read the work (in 1864, as we learn from the correspondence, as discussed below) with some care, as is evident from his extensive annotations. We may gather from these marginal comments that he found some of the theory to be compelling—particularly the Lamarckian argument that structures are adaptations that are "directly related to the differen[ces] in the incident forces [i.e., external conditions]." Darwin himself had already incorporated Lamarckian "use-inheritance" as a *part* of his own theory. At the same time, however, it is "hard to distinguish [this cause of variation] from selective spontaneous variation" (*Marginalia* 771; cf. *Variation of Animals and Plants under Domestication* vol. 2, 270–1). These comments suggest that Darwin did not find Spencer's attempt to reduce *all* variation to functional adaptation to be successful; "so-called spontaneous variation" in Darwin's sense of "cause unknown" was still in play. One may infer that Darwin continued believing that the causes of some variations really are unknown, despite Spencer's claim that all the causes are known. But Spencer's passages employing the *expression* "spontaneous variation" and Darwin's scoring of and comments on some of them make it more difficult to rule out Spencer as Darwin's source.

More light is shed on Darwin's opinion about Spencer's theory in the correspondence. In letters that Darwin sent to Spencer himself he was always nothing but cordial and amicable. He thanked Spencer in 1856 for the "extremely kind present" of the *Principles of Psychology* and looked forward to reading it (Darwin eventually did read it: *CCD* vol. 6, 56–7 [March 11, 1856] and n. 2). He wrote again in 1858 to thank Spencer for the "very kind present" of the *Essays*, in which he singled out the "development hypothesis" for special praise:

> Your remarks on the general argument of the so-called Development Theory seem to me admirable.... I treat the [same] subject [in my species book] simply as a naturalist & not from a general point of view; otherwise, in my opinion,

> your argument could not have been improved on & might have been quoted by me with great advantage (*CCD* vol. 7, 210 [November 25, 1858]).

Appreciation for Spencer's support of Darwin's theory and praise of Spencer's work on population continued in a letter from Darwin of February 23, 1860:

> I write one line to thank you much for your note. Of my numerous (Private) critics, you are almost the only one who has put the philosophy of the argument [of the *Origin*], as it seems to me, in a fair way.—Namely, as an hypothesis (with some innate probability as it seems to me) which explains several groups of facts.—You put the case of selection in your Pamphlet on population in a very striking & clear manner (*CCD* vol. 8, 105–6 [February 23, 1860]; cf. also vol. 14, 470 [December 9, 1867] for another letter of thanks for a "present" from Spencer]).

In view of Darwin's strongly positive comments about Spencer's various writings in these letters it is somewhat surprising that Darwin did *not* quote his opinions or rely on his views more extensively in the *Origin* or elsewhere. As he said to Spencer in 1858, "[your] argument could not have been improved upon & *might have been* quoted by me with great advantage" (*CCD* vol. 7, 210 [November 25, 1858], emphasis supplied). Well, why was it not? This was almost a year before Darwin had completed the *Origin* and so Spencer could easily have been quoted in that volume. True, Darwin did insert a short paragraph acknowledging Spencer in the third English edition of the *Origin* (in the "Historical Sketch") and also added three sentences to the sixth edition in which some of Spencer's views about development and psychology are briefly paraphrased (*Origin* 757: XIV. *256:f*). But in fact Darwin was always stinting in his public praise of Spencer. One surmises why this was so when one looks at what Darwin said privately about Spencer to his friends, especially J. D. Hooker, during the time when the *Biology* began to appear in print.

To judge from these letters, Darwin did not deeply admire Spencer's ideas, and neither did Hooker. Darwin asked Hooker his opinion about Spencer in January 1864 (*CCD* vol. 12 14 [January 10 and 12, 1864]), claiming not to understand him, but Hooker was the one to get the ball of serious criticism rolling. In January 1864, shortly after Spencer's *Principles of Biology* first began to appear, Hooker wrote:

> You ask about H Spencer's works, I cannot appreciate them so highly as Huxley, they are too purely speculative for me. I wonder at & worship the man's astonishing power of assimilation & incomparable fluidity of diction: [but] what I dislike most is the assumption of finality he claims, for

> all his speculations: or rather his treating all his speculative conclusions as realized facts. I cannot think him deep, but *very* ingenious, & very voluble.... I totally dispute his reasoning regarding *induced* modifications being transmitted, & think them weak & inconclusive—without however denying the fact of such heredity.—The man is I think often out of his depth. I believe he is very poor & makes his bread by these books (*CCD* vol. 12, 28 [January 24, 1864], Hooker's emphasis).

And in the same vein later that year Hooker wrote:

> I always admire [Spencer's] wonderful grasp & admirable illustrations, but his whole work is cumbrous to my mind, & reminds me of a huge mill-sluice of scientific diction & ideas: fluid, very noisy, the noise never discordant; the stream full & powerful, but never adding an inch to the depth of the river it pours into. Much of it seems to consist in clothing biological science in the language of physical science (*CCD* vol. 12, 382 [October 26, 1864]).

Darwin essentially agreed with Hooker's assessment:

> I am quite delighted with what you say about H. Spencer's book [viz., *Principles of Biology*]; when I finish each number I say to myself what an awfully clever fellow he is, but when I ask myself what I have learnt, it is just nothing.... I do not admire so much his style, & I think invariably 2 or 3 pages might be condensed into one (*CCD* vol. 12, 391 [November 3, 1864]).

While claiming to admire Spencer's "prodigality of original thought," Darwin confessed to being especially troubled by his speculations lacking any firm "scientific," that is to say empirical, foundation. Specifically, as he had already noted in his marginal comments to the *Principles of Biology*, Spencer gave no satisfactory explanation about where "the direct action of external circumstances begins and ends" (*CCD* vol. 14, 223 [June 30, 1866]). After Darwin completed reading the last installment of the *Biology* in December 1866, he continued to voice the same criticism that Spencer is too "speculative":

> I have now read the last no. of H. Spencer: I do not know whether to think it better than the previous number; but it is wonderfully clever & I daresay mostly true. I feel rather mean when I read him; I could bear & rather enjoy feeling that he was twice as ingenious & clever as myself, but when I feel that he is about a dozen times my superior, even in the master art of wriggling, I feel

> aggrieved. If he had trained himself to observe more, even if at the expense, by the law of balancement, of some loss of thinking power, hew[d]. have been a wonderful man. (*CCD* vol. 14, 427 [December 10, 1866]).

Discussion between Darwin and Hooker about Spencer's *Principles of Biology* tapers off after 1866, as does almost all written communication between Darwin and Spencer. Because none of the letters Darwin wrote to Hooker or Spencer specifically address Spencer's interpretation of "spontaneous variation," it is difficult to say, independently of the other evidence examined, that Spencer was Darwin's source for the phrase. What is clear is that Darwin believed that Spencer was a clever writer and an ingenious thinker but not much of a scientist. If one believes that Darwin would not borrow ideas or even phrases from sources he deemed to lack adequate scientific foundation, then one should conclude Darwin took "spontaneous variation" from someone else—or invented it himself. On the other hand, the similarities we have observed between Darwin and Spencer about the meaning and use of the expression, perhaps especially their joint employment of quotation marks to surround it and the phrase "so-called" to modify it, and also the timing of its occurrence in the various writings of the two men, require that Spencer remain a strong contender as the originator of the expression as used by Darwin.

A final judgment about Spencer's influence on Darwin's mode of expression is only speculation. Many pieces of evidence point to Spencer, as shown. But if we are to rely on "meaning" more than "expression" for detecting Darwin's source for "spontaneous variation," Spencer loses points. His meaning is Lamarckian "use-inheritance," wherein "functional adaptations" ("use" or "habit" in Lamarck's language) accounts for most or all variations (R. J. Richards, 1989, Chapter 6). As we have seen earlier, and will see again in the next chapter, this is not where Darwin stood. It cannot be said that Darwin failed to read Spencer with accurate understanding. But it is possible that if Darwin was influenced by Spencer to insert "so-called spontaneous variation" into later editions of the *Origin* in place of "chance" (as he was by adopting Spencer's phrase "survival of the fittest" as a substitute for "natural selection"), it may have been an influence he felt from Spencer's earlier work, not the later works in which Spencer came out as a full-fledged Lamarckian.

ARISTOTLE

Despite having reminded himself in 1838 to "read Aristotle to see if my ideas very ancient," Darwin did not get around to redeeming this pledge prior to writing the *Origin*. His familiarity with Aristotle, to the extent he had any familiarity at all, would have come through references to the Greek philosopher in other writings that he did read, but even these encounters had little discernable impact. That was prior to the appearance of the first edition of the *Origin*. In the fourth edition (1866), however, Darwin added a lengthy footnote to Aristotle. He was prompted to do this by an early reader of the *Origin*, Mr. Clair Grece, who had brought Darwin's attention to a passage from Aristotle's *Physics* bearing on the origin of species.[19]

> You may recollect me as having some year or two since pointed out to you a passage from Aristotle, shewing that "Natural Selection" was known to the ancients (*CCD* vol. 14, 386 [November 12, 1866]).

The date of this letter, and its reference to an earlier letter written "some year or two since" (the letter has not been found), shows that Darwin could have become acquainted with the Aristotle passage as early as 1864, well before he began to make his substitutions in the later editions of the *Origin* of "spontaneous variation" for "chance variation." Mr. Grece apparently thought the Aristotle passage sounded a lot like Darwin. After confessing that he had not known of this passage before, Darwin proceeded to quote the relevant section in the "Historical Sketch" in a lengthy footnote, and then to comment on it. He first pointed out that Aristotle had argued "the rain does not fall in order to make the corn grow [or] to spoil [it]," then went on to quote directly from Aristotle's text as translated by Grece:

> [Aristotle applies the same argument to organization]...So what renders the different parts [of the body] from having this merely accidental relation in nature? As the teeth, for example, grow by necessity, the front ones sharp, adapted for dividing, and the grinders flat, and serviceable for masticating the food; since they were not made for the sake of this, but it was the result of accident.... Wheresoever, therefore, all things together (that is, all the parts of one whole) happened like as if they were made for the sake of something, these were preserved, having been appropriately constituted by an internal spontaneity; and whatsoever things were not thus constituted, perished, and still perish (translated by Mr. Clair Grece, quoted by Darwin, *Origin* 59–60: "Historical Sketch." *6x. 1–4:d*).

Darwin apparently did not understand that the sentiment contained in the quoted passage was not Aristotle's, but was rather Aristotle's attempt to explain the views of another Greek philosopher, Empedocles. Aristotle went on to explain that Empedocles was wrong about "accident" as the source of variations that were useful for survival. For Aristotle the adaptive features of organisms were not accidental, but rather were brought about by "nature" to fulfill "final causes"; there was, for

Aristotle, nothing fortuitous about these structures. Grece also did not focus on the role of accident. Instead he thought the Aristotelian passage showed that "natural selection was known to the ancients." Yet those points are of little moment here. Darwin had the passage before him, and it plainly states that an argument exists, whoever its author, that adaptations are "the result of accident," that they arise by "an internal spontaneity." The meaning, judging from this passage out of its larger context, seems plainly to be a foreshadowing of Darwin's idea of "chance variation." We err, Aristotle seems to be saying, if we think that the organization of nature is built "for the sake of" some end—e.g., the teeth for the sake of dividing and chewing. Rather, the teeth were made "by accident" or "spontaneously" in a particular way, and *if* they happened to be well suited to the survival needs of the organism to which they belong, the organism is preserved; if not, it perishes.

Whether Mr. Grece recognized it or not, this *does* sound a lot like the core idea of chance variation in Darwin's theory. Darwin, in his footnote, ignored that point and decided to underscore a different one, namely, that Aristotle had not fully understood the agency of natural selection in the process of organic change, the point Grece had made in his letter to Darwin:

> We here see the principle of natural selection shadowed forth, but how little Aristotle fully comprehended the principle, is shown by his remarks on the formation of teeth (*Origin* 59–60: "Historical Sketch": *6.x*. 1–4d*).

Darwin's focus on natural selection rather than "accidental variation," in separating himself from Aristotle, may be seen as part of the larger strategy in later editions of the *Origin* to banish or obscure the role of "chance" or "accident" as much as possible from an active role in the theory. It may therefore be no coincidence that the Aristotelian phrase "internal spontaneity" occurs for the first time in the same edition of the *Origin*, the fourth, as Darwin's own rendering of "chance variation" as "spontaneous variation." It is thus plausible to surmise that Darwin took the phrase directly from Grece's translation of Aristotle.

But Aristotle also used the word "accident" in his passage, and this was a word with which Darwin by now had grown increasingly uncomfortable, at least in his published works. Thus, if Darwin did borrow from Aristotle, his dismissal of him seems somewhat disingenuous. While asserting, on the one hand, that Aristotle showed little comprehension of the principle of natural selection, Darwin, on the other, retained the most distinctive feature of Aristotle's account—chance variation—but now under the phrase "spontaneous variation" that he may have borrowed from Aristotle himself, all the while denying that Aristotle got it right!

All of this leaves unanswered the question why the teeth or any other organs are as they are, that is, what causes spontaneous variation. In his response to Aristotle Darwin did not say. He was certainly no teleologist (an important qualification in Lennox, 1993). Unlike Aristotle, he had no room for a directly intervening intelligent nature acting by foresight in his theory, any more than for an intervening deity constantly making new creations. But if nature evolves neither through chance nor through intelligence, what is left? Perhaps just what Darwin now began to say with increased frequency: "spontaneous variation."

CONCLUSION

Whatever Darwin's source for the expression "so-called spontaneous variation" may have been, two points stand out. The first is that he substituted this expression for the earlier "chance" or "accidental" variation starting with the fourth (1868) edition of *Origin*; and gave it pride of place in the sixth (1872) edition, thereby virtually eliminating any reference to or apparent role for "chance." One surmises the changes were not just "accidental." He evidently wanted to remove a noxious word and found a suitable substitute in the word "spontaneous."

One wonders whether the new word was in fact a suitable substitute. As indicated earlier, "spontaneous" could well have been seen by readers of *Origin* as less threatening than "chance" or "accident," especially for the theologically disposed. But did it mean the same thing? Two of the three authors we have considered as possible sources for Darwin—Lucas and Spencer—did not regard "spontaneity" in the same sense Darwin did. For both of them spontaneity did not mean "fortuitous" but only "causes obscure." Both also believed "obscure causes" did not mean "unknown" or even "unknowable." Spencer especially was confident that he knew what the "obscure" causes of variation were, and faulted Darwin for not seeing them (what he called "functional adaptations" to maintain "cosmic equilibrium"). Even Lucas, while invoking "spontaneity" in the appearance of new varieties, ultimately fell back on essentially Lamarckian ideas—adaptation of living creatures to new environments. Thus, in terms of *meaning,* neither Spencer nor Lucas was in Darwin's camp.

Aristotle was. Or, more precisely, Aristotle's source, Empedocles, was. Empedocles (and other naturalists of his day) did believe that variations were thrown up "by accident" and that favorable accidents would succeed in a struggle for life. Darwin did not know about these ancient theories in 1859 when he published *Origin,* so we need not think he got the *idea* of chance variation from ancient sources. But as to the expression "spontaneous variation," he may have been influenced by Grece's translation of Aristotle's *Physics* that he sent to Darwin in 1864, the first year that Darwin started to substitute in the *Origin* the word "spontaneous" for "chance" in describing how variations come about. In terms of meaning, therefore, Aristotle's rendition of Empedocles as presented in Grece's translation is a good candidate for Darwin's source.

But I think—and this is mostly conjecture—this conclusion will not do. My surmise is that Spencer was the source. The reason for thinking so is that Spencer put the expression "spontaneous" in quotes, just as Darwin did; and Spencer often prefaced the expression with the qualifier "so-called," just as Darwin usually did. The timing of Spencer's writing on this subject where he employed the expression is perfect in terms of when Darwin first started to employ it (1864). And Darwin would not have been able to see how Spencer differed from him on the origin of species (in terms of Spencer's published writings) until well after Darwin decided to adopt the phrase himself. Finally, Darwin could console himself with the thought that he was not really borrowing from someone else at all, because Spencer had himself already given Darwin "credit" for the idea of "spontaneous variation."

If this is correct, we find Spencer to have been more important in the development of Darwin's expositions of his theory than perhaps anyone else. Through Wallace Darwin received the idea to transform "natural selection" into the ubiquitous phrase (in *Origin* edition 4 and ever since) "survival of the fittest," invented by Spencer. And now, added to that well-known piece of historical investigation, we may add that Darwin's "so-called spontaneous variation" also probably owes a debt to the very same person. Considering Darwin's private contempt of Spencer's ideas, his borrowings seem less than generous, especially for one who generally accorded more than stinting praise to those who most influenced him.

NOTES

1. Darwin had already incorporated expressions like "what we may call chance," or "so-called accident" in earlier editions of the *Origin*; what is new in the fourth and subsequent editions is the expression "spontaneous variation."

2. If Aristotle was the original source, it would have been an indirect influence on Darwin, for Darwin did not read much if any Aristotle until after the *Origin* was published in 1859. The issue has been examined by A. Gotthelf (1999), *JHB* vol. 32, 3–30.

3. Cf. M. J. S. Hodge and D. Kohn in D. Kohn, ed., *The Darwinian Heritage* (1985), Princeton, NJ: Princeton University Press, chapter 6; and chapters 2–5 above.

4. Darwin may not have noticed Mill, but others did. For example, John Morley in an 1871 letter to Darwin observed: "I don't think Mr. Mill's expression in pp. 45 and 46 point to any fundamental difference between him and yourself. He admits that the moral faculty is capable of springing up 'spontaneously' in a 'certain small degree'" (*CCD* vol. 19, 302). No record exists of Darwin's reply to this letter.

5. *Variation* 9, 39, 40, 44, 48, 54–5, 65, 69, and often; *Descent* 438, 547. Darwin also had mentioned with praise Lucas's work in the *Origin*, every edition: "Any variation which is not

inherited is unimportant for us. But the number and diversity of inheritable deviations of structure, both those of slight and those of considerable physiological importance, are endless. Dr. Prosper Lucas's treatise, in two large volumes, is the fullest and the best on this subject" (*Variorum*, 85). Lucas's influence on Darwin has been examined in Noguera-Solano and Ruiz-Gutierrez, *JHB*, 2009, but without reference to spontaneous variation.

6. In view of Lucas's interest in the development of human moral faculties I find it surprising that Richards (1989, 196) does not mention him as an authority for Darwin.

7. In his *Reading Notebook* Darwin recorded reading this work on four separate occasions between 1838 and 1856 (Prichard's work first appeared in 1813), once even reminding himself "must study." His copy of Prichard (1841–1851) in the Darwin Library is annotated (*Marginalia* 683–6). Darwin's entries on Prichard in his *Reading Notebook* are recorded in *CCD* vol. 4, 437 (119: 1v); 458 (119: 6a); 475 (119: 20a); 494 (128: 18).

8. P. Lucas, (1850) vol. 2, 457–8.

9. "Spontaneous variation" differs from "spontaneous generation" insofar as it presupposes already existing life forms upon which to induce variations. "Spontaneous generation," by contrast, is a theory about the very origin of life itself. Darwin did not speculate about that in his published works, and indeed explicitly denied that his theory addressed that question. But it was possible to confuse the two ideas, as did, for example, several reviewers of Spencer's *Principles of Biology* (and to which Spencer replied in an Appendix to a later edition of his book *Principles*, 479–81). The source of the confusion is not hard to trace. If one accepts "spontaneity" in the case of variations of existing creatures, why not accept it in the case of the very origin of life? Darwin managed to evade that question, but Spencer took it head-on.

10. Darwin received a copy of this publication probably in February 1860, directly from Spencer, and wrote to Spencer expressing his thanks (*CCD* vol. 8, 105–6 [February 8, 1860]). In his letter Darwin praised Spencer's "striking and clear" argument for selection, but also, somewhat in contradiction to that statement, confessed that he had not yet read the essay, being too weak and consumed with other preoccupations to read other works.

11. In a letter from Spencer to a friend (February 10, 1860) he stated: "I am just reading Darwin's book [*Origin*] and want to send him the 'Population' to show him how thoroughly his argument harmonizes with that I have used at the close of that essay," thus again suggesting Spencer believed he had anticipated Darwin on the notion of "struggle for survival" (quoted in *CCD* vol. 8, 66, n. 6).

12. It was Wallace, not Spencer, who is immediately responsible for Darwin's making this substitution (*CCD* vol. 14, 229 [July 2, 1866]); but Wallace learned the expression from Spencer.

13. H. Spencer, 1858–1863. *Essays: Scientific, Political, and Speculative.* 2 vols. London: Longman, Brown, Green, Longmans, and Roberts.

14. It is interesting that Darwin would have chosen to mention the "Development Hypothesis" rather than the "Theory of Population" in the "Historical Sketch." But, no doubt as far as Darwin was concerned, Malthus and de Candolle had anticipated Spencer's population theory (neither of those men were entered into the "Historical Sketch" either), whereas Spencer's importance lay in his observation of "gradual modification" (i.e., the theory outlined in the "Development Hypothesis").

15. Nor does it appear in another large work that Spencer had written prior to the *Origin*, the *Principles of Psychology*, which had appeared in 1855 and that Darwin had received from Spencer in early 1856 (*CCD* vol. 6, 56 [March 11, 1856] letter from Darwin to Spencer). Darwin confessed

to Spencer in a letter on February 2, 1860 that he had not yet read this larger work, but it is interesting to observe Spencer's opinions about "modification" and "variation" at that time. He suggested that modifications in intellect could be thought of as "spontaneous," but by this he meant that such modifications were brought about by an inner principle of growth within the brain, in contrast to the Lockean notion that mental changes result from exposure to external conditions. By contrast, "variations" from generation to generation result from the Lamarckian factors of the actions on organisms of external conditions and the inheritance of characters thus acquired. In other words, variations in the 1855 work are distinctly *not* spontaneous. It again appears that Darwin's *Origin* appears to have been the mind-changer for Spencer about the cause of at least some variation.

16. Relevant passages are to be found in *Principles* 167, 246, 248, 254, 260, 264, 267, 272, 425, 451, and in an Appendix written in 1865, 480.

17. A fascinating possibility about lines of influence on Darwin is that A. R. Wallace may have been an intermediary between Spencer and Darwin. "Wallace was a Spencer enthusiast," as R. J. Richards notes (1989, 165), and was also in frequent correspondence with Darwin in the years just before and after the first appearance of the *Origin*. But I have not been able to trace any employment by Wallace of the phrase "spontaneous variation" in his correspondence with Darwin during this period.

18. See chapter 2 of this volume. All variations, Darwin was sure, are caused. Where he departed from Spencer was in his conviction that, whatever the cause(s) may be, they (or some of them) are at present *not* known. Spencer's account did not satisfy him. Nor would he accept accounts that put the cause of variation in a divine intelligence. His skepticism about Spencer's account would not have troubled many people, even if they had known about it (the only one in whom he confided about Spencer's deficiencies as a thinker was Hooker, and Hooker largely shared Darwin's skepticism). But Darwin's refusal to accept Divine intelligence, as shown above, was troublesome to many, and Darwin's dodge in saying "so-called spontaneous variation" is only a placeholder, as "our ignorance" did not mollify them.

19. Discussions include A. Gotthelf *JHB* 1999, 3–22; C. Johnson, *JHB* 2007, 529–56. David Keyt (1987) discusses "spontaneity" as a "cause" (among four) of how things or events come to be in Aristotle's thought more generally (his translation of Aristotle's *automaton* in several Aristotelian works), but it is virtually certain Darwin had no familiarity with Aristotle's works cited here or with much if any Greek.

7

Darwin's Architect Metaphor

THE PUBLICATION OF Darwin's masterpiece *Origin of Species* in 1859 caused an immediate stir in Britain both in scientific and non-scientific circles. Criticisms, often hostile, poured in from both quarters. Some of it came from people who understood that he was denying "special creation" and believed this rendered his theory not just unacceptable but downright odious. Others simply misunderstood: they thought Darwin had saved a place for special or "directed" creation and faulted him for not making its place in his theory clearer, more explicit.[1] (Few early critics grasped that Darwin was denying

YBP Library Services

JOHNSON, CURTIS N., 1948-

DARWIN'S DICE: THE IDEA OF CHANCE IN THE THOUGHT
OF CHARLES DARWIN.
Cloth 253 P.
NEW YORK: OXFORD UNIVERSITY PRESS, 2015

EXAMINES WHAT DARWIN MAY HAVE THOUGHT ABOUT
VARIATIONS & CHANCE REQUIRED FOR NATURAL SELECTION
LCCN 2014010297
ISBN 019936141X Library PO# FIRM ORDERS
List 29.95 USD
8395 NATIONAL UNIVERSITY LIBRAR Disc 14.0%
App. Date 4/22/15 COLS-SCI 8214-08 Net 25.76 USD

SUBJ: 1. DARWIN, CHARLES, 1809-1882--PSYCH. 2.
SERENDIPITY IN SCIENCE.

CLASS QH365 DEWEY# 576.82 LEVEL ADV-AC

intelligence *and* were comfortable with that stance.) To meet critics of both sorts Darwin fashioned a "stone house" or "architect" metaphor, in 1860.[2] The metaphor, he believed, got to some core issues that many of his readers seemed not to understand properly. In it he hoped to show more clearly, by analogy, how natural selection works, how it is related to variation in the production of new species, and how it removes the need for special or "directed" creation in a theory of organic evolution.[3]

But Darwin's main motivation for introducing the metaphor, first in private to select correspondents, then later in a major publication (*Variation of Animals and Plants under Domestication*, 1868), may have been to defuse criticisms of his theory coming from people who were worried about "chance" in the theory. (Such reworkings of the presentation of his theory over many years is the central theme of this book.) The metaphor retains, even strengthens, the place of chance. But it also recasts it in such a way as to make it seem less threatening to people with theological dispositions. Architects (in the metaphor) employ stones that fall "randomly" from precipices to construct their buildings, but we need not suppose that "God" or an intelligent designer caused the stones to assume the shapes they do just to suit the architect. In the same way, by analogy, we need not suppose God ordained the specific variations exhibited in natural organisms just to suit breeders' fancies (in the case of domesticated productions) or even to suit the survival needs of organisms in the state of nature. If fallen stones at the base of a precipice need occasion no alarm about a Godless world, neither should "random" variations. The two cases are instances of the same type.[4]

However, Darwin's choice of the architect metaphor may have worked directly at cross-purposes with his intentions. Architects, unlike natural selection, *are* intelligent.[5] They make conscious choices in directing the creation of their new productions. In any case, Darwin's choice of this image was peculiar. The image for centuries had been frequently employed to connote the very deity that Darwin was trying to exclude.[6] But it did soften the blow. It did so not by allowing a role for intelligence in natural selection—that was not what Darwin wanted to convey—but rather by showing that chance in variation is not such a dangerous idea as people might have come to think.[7]

"Chance" enters Darwin's theory at the point of variation, the "raw material" upon which natural selection works.[8] If variations are not designed, how do they come about? Darwin's belief was that they are "caused" by the complex action of several "natural laws," some of which are only "dimly understood" or even not understood at all. When Darwin did not understand why a particular variation arose, he often, at least in the first edition of the *Origin*, referred to the modification as having occurred "by chance." He quickly came to see, however, that this expression was bringing grief to some of his critics, even friendly ones such as Charles Lyell and Asa

Gray.[9] For men such as these, the affirmation of chance implied the denial of design, and Darwin was making matters worse by saying so explicitly.[10]

To counter this reaction Darwin inserted a statement at the beginning of Chapter Five of *Origin*, "The Laws of Variation," that insisted the phrase "chance variation" is a wholly incorrect statement, used only as a cover for "our ignorance" (*Origin, Variorum* edition, V, 5[a], 275).[11] That demurrer, though, was not enough. He continued to be understood to be saying that variations are "by chance" or "accidental." Part of his strategy to address this ongoing concern was silently but systematically to expunge the words "accident" and "chance" (as applied to variation) from later editions of the *Origin* (Johnson, 2010). These changes were not intended by Darwin to bring him into alignment with his opponents who insisted on design. Rather, they seem to have been made in the hope that the criticism of his theory from "chance" might go away with the expurgation of the word.[12]

While these modifications to the *Origin* were being made, Darwin was also working on the first part of his deferred "big species book," which he took up again in 1862, after the lengthy hiatus during which he wrote the *Origin*. This latter work first appeared in published form in 1868 as *Variation of Animals and Plants under Domestication*.[13] It was only in this book that Darwin first went public with his architect metaphor. In it, the architect (natural selection) fashions noble buildings (i.e., new species) from an assortment of rocks and bricks that happen to be lying around (i.e., variations). (The metaphor did not make an appearance in any edition of the *Origin*, somewhat surprisingly.) But the metaphor had a prior history in Darwin's thought, revealed in his correspondence. What one discovers in examining this history is that Darwin made successive changes to this metaphor in parallel with changes regarding the role of chance in the *Origin*. But the changes run in the opposite direction. Instead of diminishing or at least obscuring the role of "chance" in evolution, the architect image, contrary to expectations, came over time actually to enhance it. But Darwin may also have believed that the architect metaphor might render chance less noxious to his critics. Better to be honest about it than to sweep it under the rug. As embedded in the metaphor, chance does not seem as threatening as it had before, no more dangerous than fragments of rocks at the base of a cliff. Who would worry about organic variations that were analogous to that? Nevertheless, in its final iteration the architect metaphor, fleshed out in its full detail in 1868, could only have strengthened the impression that "chance," suitably defined, does indeed lie at the base of Darwin's understanding of the natural order.

NATURAL SELECTION, VARIATION, AND CHANCE

As far can be gathered from his published writings, Darwin thought of the architect metaphor only months after the first appearance of the *Origin* in 1859.[14] He seems to have been prompted by the confusion people were having in grasping the distinction between natural selection and variation. The distinction is important and was not always clearly perceived by some of Darwin's readers and reviewers. Darwin later came to admit that he was not always as clear in expression as he should have been (e.g., *CCD* vol. 8, 496 [November 26, 1860], letter to Asa Gray). He had an extended exchange on the topic with Hooker beginning in late 1859 and continuing into the summer of 1860. Hooker had suggested that Darwin was "rid[ing] natural selection too hard," and had overlooked the role of variation in effecting "improvements" (*CCD* vol. 7, 437 [December 20, 1859]). At first Darwin replied that "selection is **the** efficient cause" in effecting the creation of new species, while also acknowledging a role for variation (*CCD* vol. 8 p. 230 [May 29 1860], emphasis in the original). He immediately rethought his wording, for only a week later he wrote again to deny that he had ever called natural selection "the efficient cause [of the origin of species] to the exclusion of the other, i.e., variability," but instead regarded it only as "an active handmaid influencing its mistress [variation] most materially" (*CCD* vol. 8, 238 [June 5, 1860]).[15]

But despite Darwin's best attempts to keep the concepts separate, some of his critics persisted in confusing variation, the source of the raw material upon which natural selection works, with natural selection itself. Charles Lyell seems to have made this error, for he wrote to Darwin in September 1860:

> Instead of Selection I should have said, Variation & Nat. Selection. My only objection is not to the term, but to your assigning it more work than it can do & the not carefully guarding against confounding it with the creative power to which "variation" & something far higher than mere variation viz. the capacity of ascending in the scale of being, must belong. Most likely you would have chosen some term less worthy of Deification [i.e., variation?], for "selection" you had an excellent technical reason (*CCD* vol. 8, 400 [September 30, 1860]).

Darwin wrote back immediately:

> Hooker made same remark that it ought to have been "Variation & nat. selection." Yet with domestic productions, when Selection is spoken of, variation is always implied.... I do **not** agree with your remark that I make N. Selection do too much work.—You will perhaps reply, that every man rides his Hobby-horse to death; & that I am in this galloping state (*CCD* vol. 8, 403 [October 3, 1860], emphasis in original).

But it is not clear that Lyell ever came to see the bright line separating variation from natural selection, this despite the fact that Darwin did his best to make the distinction as clear as possible even before he introduced the stone-house metaphor. In a letter to Lyell on September 1, 1860, Darwin had this to say:

> I have said what I can in defence [of my theory]; but yours is a good line of attack. We should, however, always remember that no change will ever be effected till a variation in habits or structure or of both *chance* to occur in right direction so as to give the organism in question an advantage over other already established inhabitants of land or water; & this may be in any particular case indefinitely long (*CCD* vol. 8 [Sept. 1, 1860], 339, original emphasis).

Despite clarifications such as this, Lyell still, as late as 1865, seemed to think the two words—variation and natural selection—referred to much the same phenomenon, both to be distinguished from the really important role of a "creative power" or "creational law" in evolution: "variation or natural selection cannot be confounded with the creational law" (*CCD* vol. 13, 22 [January 16, 1865], to Darwin from Lyell).[16]

One sees evident confusion on both sides—Lyell's and Darwin's. Lyell seems to conflate natural selection and variation into a single concept (note his choice of the conjunction "or" instead of "and" to refer to the two-step Darwinian process). But Darwin may have misread Lyell's letter. What term was Lyell complaining about when he said Darwin should have chosen a term more worthy of "deification": natural selection or variation? Darwin took him to mean "natural selection" as the "hobby-horse" he was riding too hard. But Lyell's statement in the September 30, 1860, letter (quoted above) seems to be referring to variation. Lyell did accept natural selection as a proper explanation for species change. What he objected to was "unguided variation." Darwin's formulation, "variation *and* natural selection" made no reference to the creative force Lyell cared about most. But, whatever the correct interpretation of Lyell's meaning is, both Darwin and Lyell settled into a mode of debate in which the meaning of Lyell's term "deification" was never clearly spelled out. Darwin seemed always to think it referred to Natural Selection, whereas Lyell apparently intended it to mean "variation or natural selection."

One cannot blame only Lyell for confusion. Hugh Falconer, a distinguished paleontologist and botanist who had studied *Origin* with some care, nevertheless concluded in January 1863:

> The means which have been adduced to explain the origin of species by "Natural Selection" or a process of variation from external influences, is inadequate... it is difficult to believe that there is not in nature, a deeper seated and innate principle, to the operation of which "Natural Selection" is merely an adjunct (quoted in *CCD* vol. 11, 15, n. 6).[17]

Hooker too saw Falconer's evident confusion and pointed it out to Darwin: "[Falconer] too [i.e., like Lyell?] thinks that you make Nat. Selection work independently & do everything without variation to work upon" (*CCD* vol. 11, 14 [January 6, 1863]). Darwin's frustration is evident in his reply to Hooker a week later: "How any man can persuade himself that species change unless he sees how they become adapted to their conditions [through natural selection] is to me incomprehensible" (*CCD* vol. 11, 36 [January 13, 1863]).

The same sort of misunderstanding shows up in Darwin's correspondence with W. H. Harvey in mid-1860: "It is an extraordinary fact that he [Harvey] does not understand at all what I mean by Nat. Selection," Darwin lamented to Asa Gray in September (*CCD* vol. 8, 389 [September 26, 1860]). What he meant by this is explained in a letter from Darwin to Harvey himself some days earlier: "You speak...as if I had said that natural Selection was the sole agency of modification; whereas I have over & over again, ad nauseam, directly said and by order of precedence implied (what seems to me obvious) that selection can do nothing without previous variability" (*CCD* vol. 8, 371 [September 20–24, 1860]).[18]

It is surprising that such confusion should reign among a scientific audience on such a pivotal issue in Darwin's theory. But the issue is in fact complicated. One source of the confusion is a question that Darwin was raising about what *causes* modification of species. Two answers are in play because two factors are in play—variation and natural selection. Thus, Darwin was really asking two questions almost simultaneously: what causes modification of species (or populations), and what causes variation among individuals. Usually Darwin was focused on "natural selection" as the *primary* (though not the only) cause of modification of species. But natural selection cannot act without something to act upon—actual organisms in all of their variety. It is thus a short mental step to steer the question to the causes of variation.

On that subject Darwin's thinking was more complicated. Variation, he believed, had multiple causes: "direct action of conditions on organization" (the "Geoffroyian" explanation); use/disuse and habit (the "Lamarckian" explanation); the indirect action of physical and organic conditions on the reproductive system of organisms; correlations of growth (not really "causes," but still "laws"); and some "unknown" cause that he usually referred to as "chance" (see chapter 3 for details). Often his critics simply did not see the difference between the two kinds of questions, nor did they see Darwin's attempts to discriminate among the various "causes" that he assigned to the second factor, variation.

Further complicating matters is that some of these same critics, especially Lyell and Asa Gray, believed that yet another cause of variation existed, perhaps better called a meta-cause standing behind and guiding the other causes: a "creative power" or "creational law." Lyell is typical in this regard: on one side is "variation or natural selection," as if the two were interchangeable terms, on the other side some "higher power" or "creational law" that guides the whole process of variation at every stage. Lyell thought that Darwin had assigned too great a role to "natural selection or variation" at the expense of this higher cause; indeed, that was his chief complaint with Darwin's entire theory. Darwin insisted that his theory did not require this "higher power," and that if it did he would "reject it as rubbish" (*CCD* vol. 7, 345 [October 11, 1859], to Lyell).

Variation does come first in the origin of new species, but what causes variation is not to be understood as a special and mysterious "creative power." True, the causes of variation are complicated and often "unknown," but Darwin could not accept a creative or higher power. That is really the root of the difference between Darwin and many of his critics that lies under the surface of the misunderstandings that we have witnessed above. Lyell, Falconer, Harvey, Kingsley, Gray—and no doubt many others—wanted a higher power in variation, and Darwin did not. Darwin's answer to these men was to introduce the architect metaphor that he believed explained things more clearly.

DISCOVERY OF THE ARCHITECT METAPHOR

The first reference to the architect image that I have been able to trace comes in a letter written by Darwin to Hooker in June 1860:

> The following metaphor gives good view of my notion of the relative importance of Variation & Selection.—Squared stones, bricks or timber are indispensable for the construction of a building; & their nature will to certain extent influence character of building, but selection I look at, as the architect; and in admiring a well-contrived or splendid building one speaks of the architect alone & not the brick maker (*CCD* vol. 8, 252 [June 12, 1860]).

Darwin was pleased with the analogy, for he repeated it to Lyell two days later:

> The more I study the more I am led to think that natural selection regulates in a state of nature most trifling differences.—As squared stones, or bricks, or timber are the indispensable materials for a building & influence its character; so is variability not only indispensable, but influential; Yet, in same manner, as the architect is the *all*-important person in a Building, so is Selection with organic bodies (*CCD* vol. 8, 254 [June 14, 1860], emphasis in original).

Darwin was attempting to fend off the criticism that he had made natural selection "do too much work," and that he had not allowed sufficiently for the role of other factors in producing organic change. The idea the image conveys is that two influences are at work in the origin of new species: variations (the stones, bricks, timber—i.e., raw materials), and "selection," the picking out of appropriate materials and fashioning them into something new and splendid—new species.

In a way Darwin was addressing the wrong question, for most of his critics on this score were not so worried about the distinction between natural selection and variation as they were about the absence in Darwin's theory of any "higher power" or "creational law" that would account for variations. Darwin's metaphor to this point did not address that concern. In fact, it misleadingly implies two things that Darwin certainly did not mean to imply: first, that the stones and bricks (although not necessarily the timber)—and so, by analogy, the variations—are themselves "designed" (after all, one does not find squared stones or bricks spontaneously arising in nature, and Darwin explicitly mentioned their "maker" in the first passage above!); and that, just as the architect makes intelligent choices about how to put these materials together to fashion buildings, so too, by analogy, selection acts intelligently to bring forth new organic productions. Darwin seemed to be admitting entrance at two points into his theory of the very intelligent design that he had hitherto tried hard to exclude.

Lyell was quick to pick up on the second point. In a short letter to Darwin written apparently the very day he heard from Darwin about the architecture metaphor he said he could not accept it as an appropriate metaphor for how new species are produced. His objection was not to the idea of design implied in the "squared stones and brick," for he did not mention this factor. Instead he complained about the comparison of selection to an architect. The architect's work, Lyell suggested, is far more complicated and intelligent than "the humble office of the most sagacious of breeders," let alone, presumably, the "blind" action of natural selection. Once again, Lyell was worrying that Darwin's theory of "variation or natural selection" eliminated any need for a creative power. He just could not go along with this.

Lyell's criticism, however, is ambiguous. It could mean that Darwin was assigning "too much weight" to selection, vis-à-vis other factors. Or, it could mean that Darwin was assigning to it some power of intelligence—as architects in fact exhibit when they fashion raw materials into beautiful buildings. The latter reading is at first more plausible. When he first objected to the architect metaphor Lyell drew special attention to the thinking, planning, and intelligence that go into the architect's craft. The analogy thus suggests a thinking, planning, intelligent natural selection as the corresponding idea to the architect in the metaphor.

But that interpretation fails. For one thing, if Darwin had meant to imply that natural selection is intelligent one might have expected Lyell to have given the image a better reception. After all, what always bothered Lyell about Darwin's theory was that it seemed to eliminate intelligence from natural productions altogether. Wouldn't a little "intelligence of the architect of nature" be just the sort of concession to design that Lyell had hoped for? In addition, one sees from subsequent letters that Darwin did not read Lyell's objection in that way, and that Lyell did not intend it in that way. Instead, both men understood Lyell to be saying that natural selection was being assigned "too much weight." Greater weight, Lyell thought, needs to be assigned to the "raw materials" of selection, and more than that, to a "creative force" that produces and guides variations. Lyell was objecting to Darwin's *elimination* of the deity, not just the "deification" of variation or natural selection.

Darwin immediately wrote back to give Lyell reassurance: to say that natural selection has "so much weight" in effecting organic change is not to rule out a role for other factors. He ignored what could have been taken to be Lyell's main point, that to make natural selection an "architect" was tantamount to making it intelligent. The sensible answer to this would be simply to say the architect metaphor was just that, a metaphor, and like other metaphors it could be misleading in some parts. This was just the strategy Darwin employed against other critics who accused him of personifying natural selection. But instead, to Lyell Darwin replied that just because natural selection was important, even "all important," in producing species does not mean that it acts alone. Not only must one recall the importance of variations, but even beyond that, Darwin now insisted, one must also keep in view the "more general laws" that govern the universe in all of its workings:

> The very existence of the architect shows the existence of more general laws; but no one in giving credit for a building to the human architect thinks it necessary to refer to the laws by which man has appeared. No astronomer in showing how movements of Planets are due to gravity, thinks it necessary to say that the law of gravity was designed [so] that the planets shd. pursue the courses which they pursue. I cannot believe there is a bit more interference by the Creator in the construction of each species, than in the course of the planets (*CCD* vol. 8, 258 [June 17, 1860]).

Darwin's response does not affirm or deny that, since the architect is intelligent, so is natural selection. At best he implies that it is *not* intelligent by suggesting that it operates like other "general laws." This letter to Lyell is to be read as a defense by Darwin of his failure to bring into his discussion of the origin of species the operation of these more general laws: Darwin's theory can be presented and understood without reference to them (as could Newton's). Were these "general laws" themselves "designed?" An answer to that question may have come close to Lyell's concern, but Darwin does not quite give one. His references to "design" and the "Creator" in the passage do not say or imply that Darwin was committed to these higher powers. He only says that no reference need be made to them in an account of the origin of species. The question, he suggests, is a distraction from science. Darwin was quite sure Lyell would never agree.[19]

Darwin was right—Lyell did not agree. His response to Darwin, written two days later, and written despite the fact that Darwin had told him not to bother with a response, was essentially a defense of "Paley & Co.," that is, the attempt to show intelligent design in nature. "Constant laws," he wrote, simply cannot account for at least *some* new organic productions: the origin of life itself, the appearance of man, his instinct, reason, mind and soul, and human free will. Free will especially seems inconsistent with "constant laws," for it ensures actions will occur that are not governed by the necessity that "constant laws" implies:

> The free will of Man which however inconsistent with belief in constant laws you must admit or give up your source of all knowledge and ignore the constitution of your own mind, must I think have some counterpart in the Deity or First Cause according to the highest conception I can make of him or it. Volition or Free Will in Man is a new cause which in the time of the Deinosaurians had no action, did not interfere with the course of the vital action in the globe (*CCD* vol. 8, 260 [July 19, 1860]).

Lyell apparently was still under the impression that Darwin had "deified" natural selection, for he called the Paleyan approach of likening the Unknown Cause of new productions to the Mind and Soul of Man "far more philosophical" than the mechanistic view that "deifies Matter & Force or Natural Selection."[20]

Darwin did not budge. Again, just days later he wrote back to Lyell, implicitly defending the architect metaphor. He thanked Lyell for his patience, but could not agree that "natural selection" belonged in the same camp with Huxley's "force and matter" as adequate to explain natural processes and productions. His reason was that he did not believe "natural selection" either placed too much weight on "secondary causes," or, as Lyell had argued, placed secondary causes beyond human understanding. It would only be "beyond our depth" if it presumed to give an account of "first causes," and this it does not try to do. Often we must content ourselves with the realization that "first causes," here to be understood as the causes of variation, are simply unexplainable (*CCD* vol. 8, 262 [June 20, 1860]).

Darwin also disagreed with Lyell about free will. Lyell had made free will into a "new first cause," one that could not have been generated without contradicting the action of more "constant laws." Free will combats a deterministic universe. It enables creatures that possess it to intervene in natural processes, if not actually to subvert constant laws, at least to add to nature's complexities a "cause" of a different kind, an unconstrained and "non-constant" cause. Thus, Lyell reasoned, it must itself have had an origin in a First Cause of a different kind.

Darwin at this point could have reverted to a position he had arrived at in his earliest reflections on origins as recorded in the *Notebooks*. There he had concluded, after much reflection, that humans do not possess free will. They are governed by the same laws of necessity that govern other natural phenomena.[21] But to Lyell he took an opposite approach. The dinosaurs, he claimed, do have free will, and therefore the argument from free will in man alone loses its force (*CCD* vol. 8, 262 [June 1860]). One may only surmise what Darwin may have intended by this somewhat shocking assertion, but it is probable, in light of his other discussions about free will, that he was thinking in typical gradualist terms. Just as species throughout organic life are connected—at least in theory—by slight gradations in organic structure and function, so too no sharp breaks should be posited in mental capacities, including in the capacity for volition. All stand or fall by the same criterion. In a sense, Darwin's claim in the *Notebooks* that humans *lack* free will is just another way of saying that lower animals *possess* it. The point is that no clear line of separation between the two exists. In both cases decisions about "what to do" are present to the consciousness of the actors (men or beasts). Such decisions are guided either by reflection or by instinct, but in either case both men and dinosaurs follow the course they deem best in the circumstances. Maybe this is free will, maybe it is not. (The subject is treated at greater length in chapter 9 of this volume.)

MODIFICATIONS OF THE ARCHITECT METAPHOR 1860–1863

Lyell apparently did not write back on this subject, and Darwin too let go of the discussion. His letter to Lyell in June 1860 was the last he had to say about his architecture metaphor for a couple of years. He did pick up the theme again in 1863 with two correspondents, Asa Gray and Patrick Matthew, and again in 1868 in his correspondence with Hooker. He continued to show his belief in its illuminative powers by inserting several paragraphs that employed it in his 1868 publication *Variation of Animals and Plants Under Domestication*. But before turning to that part of the story it is worth looking again at how the metaphor played out in Darwin's correspondence in the immediate aftermath of the *Origin*. Lyell had taken the metaphor that likened variation and natural selection to random stones and an architect to "deify" the whole evolutionary process, and he objected. By this he meant that it was made to do "too much work" in effecting new productions. Lyell wanted some role for a "creative force." What is missing from Lyell's objections, and also from Darwin's responses, is any consideration of how large a role in speciation should be assigned to variation in contrast to natural selection. The two men argued about whether natural selection was carrying too heavy a load but did not get around to the question of whether variation could take on some of the baggage, and how much.

Yet Darwin's metaphor, as first described to Hooker and Lyell in 1860, seems to have been custom-made for tackling this very question. In the metaphor variation was likened to the "squared stones, brick, and timber" that the architect takes up for use in fashioning new buildings; natural selection in the image corresponds to the architect. For some reason Lyell and Darwin got caught up in debating the architect (natural selection) and ignored the "raw material" (variation). But the stones are "squared"; the bricks are ready-made; even the timber seems to be prefabricated for construction. How did these marvels arise? In his first use of the image (dropped in later renditions) Darwin claimed they had a "maker." Who might that be? Surprisingly, neither man pursued these fascinating and potentially revealing issues.

Nevertheless, Darwin in time evidently came to see the importance of these questions. The way he handled them was to change the metaphor. The first change he made was in passages we have already examined. After introducing the metaphor to Hooker in 1860 with the words, "In admiring a well-contrived or splendid building one speaks of the architect alone & not the brick maker" (*CCD* vol. 8, 252 [June 12, 1860]), he two days later in a letter to Lyell spoke only of "the architect" and "the bricks"; the "brick-maker" had fallen out (*CCD* vol. 8, 254 [June 14, 1860]). One suspects that Darwin realized the implication and so removed all mention of a maker. Two years later he made an equally significant change, this time in a letter to the American botanist Asa Gray (who had presumably not yet heard about Darwin's architecture usage in any of its forms), by replacing "squared stones, bricks, and timber" with the very different image of "stones fallen from a cliff":

> In my present book [viz., *Variation of Animals and Plants Under Domestication*][22] I have been comparing variation to the shape of stones fallen from a cliff, & natural selection or artificial selection to the architect; but I cannot at all work a metaphor like you do" (*CCD* vol. 11, 581 [August 4, 1863]).

One does not expect stones fallen from a cliff to be "squared stones" or "brick," but rather just carelessly composed fragments bearing no marks of the hands of man. Nevertheless, the architect may use those that suit his purposes, so the image of the architect still applies as before, although now with an even greater suggestion of an intelligent selector. The "architect" in the metaphor now has to pick and choose among a wide range of variant shapes and sizes to find those suitable to effect the building he intends to make. In one way the new description is an improvement: variations in organic productions do in fact present more choices for selection than the small number suggested by "squared stones, brick, and timber." But this improvement comes at the cost of sharpening the implied role of "intelligence" in selection. Darwin in fact did not intend the metaphor to suggest intelligent selection in any version of his metaphor. But unwittingly or not, that is what he got.

A further change is hinted at, but not developed, in the 1863 letter to Asa Gray. Immediately before mentioning the architect metaphor Darwin had chided Lyell for not responding to a question that much interested and impressed Darwin: how did the variations that gave rise to pouter and fan-tail pigeons arise? Darwin could not believe that these variations were designed by an intelligent agent, for to believe so would strain credulity: why would an intelligent designer cause pigeons to vary simply in order that man may select certain features to please his own fancy? But, if pigeons under domestication are not designed, why then should we think any more that variations in nature are designed? This was always one decisive "proof" for Darwin in believing in non-designed variation.[23] His alternative? As he says in his letter to Gray, "are not these variations accidental as far as the purposes of man has put them to?" So, too, one might ask, are not the fragments of stone that have "fallen from a cliff" accidental so far as the architect's purposes are concerned?[24]

Darwin further tinkered with the image later in 1863, in a letter to Patrick Matthew. Darwin had acknowledged Matthew in the "Historical Sketch," added to the third and subsequent editions of the *Origin* to give a brief "history of opinion" on the species question prior to his own contribution, as one who had anticipated Darwin's theory of natural selection.[25] The letter to Matthew made a few modifications to the architect metaphor (the letter was dictated to Darwin's wife, Emma, who transcribed as follows):

> With regard to Natural Selection he [i.e., Charles Darwin] says that he is not staggered by your striking remarks. [The letter from Matthew to which Darwin was responding has not been found.] He is more faithful to your own original child than you are yourself. He says you will understand what he means by the following metaphor:
>
> Fragments of rocks fallen from a lofty precipice assume an infinitude of shapes—these shapes being due to the nature of the rock, the law of gravity &c.—By merely selecting the well-shaped stones & rejecting the ill-shaped an architect (called Natural Selection) could make many and various noble buildings (*CCD* vol. 11, 674 [November 21, 1863]).

Both variation and natural selection are, in this new rendering, made to do even more work in bringing forth new species than before. Now the fragments, instead of being merely "stones fallen from a cliff," have assumed "an infinitude of shapes." Moreover, an account is now given of how they have assumed their various sizes and shapes in terms, presumably, of the laws of physics: "the nature of the rock, the law of gravity, &c." This is quite a long distance from the earlier image of a "maker"! Natural selection also has become more sophisticated. Instead of merely fashioning the stones into buildings, natural selection now "select[s] the well-shaped stones and reject[s] the ill-shaped [ones]." That latter change not only strengthens (yet again) the idea of intelligence in selection. It also introduces for the first time into the image the two-fold power of natural selection—its ability not only to prune and eliminate unfavorable or unfit organisms from nature's polity, but also to *advance* organic diversity by preserving and adding up novel variations to produce new species. After all, an architect is a builder much more than a grim reaper.[26]

STONE HOUSES IN *VARIATION* (1868)

Except for the suggestion that natural selection is in some sense an intelligent and creative actor, the metaphor as worked out in the letter to Matthew is just about everything Darwin could hope for. He had by now become sufficiently comfortable with it that he inserted it in 1868 into his new book *Variation of Animals and Plants under Domestication* for public inspection. He included it in three separate places in this large work, the first at the close of his chapter on "Selection By Man," the second and third at the end of the work. The placement suggests that he still regarded it as a useful and illuminating way to think about his theory. The first passage is this:

> Throughout this chapter and elsewhere I have spoken of selection as the paramount power, yet its action absolutely depends on what we in our ignorance call spontaneous or accidental variability. Let an architect be compelled to build an edifice with uncut stones, fallen from a precipice. The shape of each fragment may be called accidental; yet the shape of each has been determined by the force of gravity, the nature of the rock, and the slope of the precipice,—events and circumstances, all of which depend on natural laws; but there is no relation between these laws and the purpose for which each fragment is used by the builder. In the same manner the variations of each creature are determined by fixed and immutable laws; but these bear no relation to the living structure which is slowly built up through the power of selection, whether this be natural or artificial selection (*Variation*, vol. 2, 236).[27]

The passage emphasizes the role of "natural laws" as the cause of variations and also the lack of any relation between variations thus naturally (or "accidentally") produced and the "purposes" that these variations will serve in the polity of nature. This is a deliberate attempt to exclude "design" from Darwin's theory of variation. The second passage reaffirms the point:

> Although each modification [i.e., variation] must have its own exciting cause, and though each is subjected to law, yet we can so rarely trace the precise relation between cause and effect, that we are tempted to speak of variations as if they arose spontaneously. We may even call them accidental, but this must be only in the sense in which we say that a fragment of rock dropped from a height owes its shape to accident (*Variation*, vol. 2, 416).

What is new in these developments of Darwin's architect metaphor is an increasingly prominent role, appropriately qualified, for "accident" or "spontaneity" in causing variations.[28] Accident had been suggested in Darwin's letter to Gray in 1863, but it had not been made explicit. Now it is explicit, with the caveat that "accident" is not to be understood as "uncaused," but only as "not designed with reference to future adaptive needs." The final passage in the *Variation* goes to some length to make this distinction clear:

> I may recur to the metaphor given in a former chapter: if an architect were to rear a noble and commodious edifice, without the use of cut stone, by selecting from fragments at the base of a precipice wedge-formed stones for his arches, elongated stones for his lintels, and flat stones for his roof, we should admire his skill and regard him as the paramount power. Now, the fragments of stone, though indispensable to the architect, bear to the edifice built by him the same relation which the fluctuating variations of organic beings bear to the varied and admirable structures ultimately acquired by their modified descendents.... The shape of the fragments of stone at the base of our precipice may be called accidental, but this is not strictly correct; for the shape of each depends on a long sequence of events, all obeying natural laws; on the nature of the rocks, on the lines of deposition or cleavage, on the form of the mountain, which depends on its upheaval and subsequent denudation, and lastly on the storm or earthquake which throws down the fragments. But in regard to the use to which the fragments may be put, their shape may be strictly said to be accidental (*Variation*, vol. 2, 426–7).

In this important passage Darwin made some significant adjustments to his metaphor. He continued to imply that natural selection is intelligent: it "selects," it has skill, it has paramount power. But these are not the details Darwin was worried about. Instead he wanted to clarify his ideas about variation. The emphasis is now on their "accidental" appearance, like the shape of rocks that have fallen from a precipice. "Cut stones," as was suggested in his first use of the image, are now explicitly ruled out. They are replaced by "fragments." At the same time, however, the fragments are not exactly random occurrences.[29] They are the result of the complex interaction of numerous natural laws. The list of these laws is considerably expanded from what we have seen before: "the nature of the rocks, the lines of deposition or cleavage, the form of the mountain, which depends on its upheaval and subsequent denudation, and lastly the storm or earthquake which throws down the fragments." The overall impression generated by these changes to the architect metaphor is a further diminution of the action of any creative power in effecting organic change. Darwin was still dealing with his critics, and with characteristic doubts, spells out his stance as well as he could:

> And here we are led to face a great difficulty, in alluding to which I am aware that I am travelling beyond my proper province. An omniscient Creator must have foreseen every consequence which results from the laws imposed by Him. But can it be reasonably maintained that the Creator intentionally ordered, if we use the words in any ordinary sense, that certain fragments of rocks should assume certain shapes so that the builder might erect his edifice? If the various laws which have determined the shape of each fragment were not predetermined for the builder's sake, can it be maintained with any greater probability that He specially ordained for the sake of the breeder each of the innumerable variations in our domestic animals and plants? (*Variation*, vol. 2, 427).

What he gave with one hand ("an omniscient Creator must have foreseen every consequence") Darwin quickly took away with the other: one cannot "reasonably maintain" that a Creator intentionally designed the shape of rock fragments that have fallen off a precipice for the precise use to which the architect puts them. By extension one cannot believe that such a Being should have preordained every variation used by breeders to shape new races of organic beings, or that organic variations in nature should have been so created (*Variation*, vol. 2, 427–8). The emphasis in Darwin's image is now squarely on the "accidental" nature of variation, where "accident" means not uncaused but rather not preordained.

DARWIN AND ASA GRAY

This latest version of the metaphor shows that Darwin had come to see that his theory was more vulnerable, from the standpoint of design, to criticisms about his account of variation than to criticisms about natural selection. Natural selection seemed to be intact on this score, whereas variation needed shoring up against the notion that an omniscient deity had foreordained each new slight modification. Asa Gray, who reviewed *Variation* in the March 19, 1868, issue of the *Nation*, changed the line of attack. For the first time someone challenged Darwin's metaphor on the score of the apparent intelligence implied by the metaphor of natural selection. In his review Gray wrote:

> In Mr. Darwin's parallel [between the architect and natural selection], to meet the case in nature according to his own view of it, not only the fragments of rock (answering to variation) should fall, but the edifice (answering to natural section) should rise, irrespective of will and choice (Asa Gray, *Nation*, vol. 6, 235).[30]

Gray had hit the nail on the head:[31] if the metaphor is to hold constant with Darwin's own views, natural selection should not be allowed to choose, as an architect does. Just as the fragments fall from the precipice without design, so the edifice should rise without intelligence. Darwin, Gray seemed to be suggesting, violated his own understanding of natural processes by appearing to admit intelligent design at the point of selection. Even Gray could accept "blind necessity" for natural selection, once variations had been given. He would only have balked at the suggestion of "pure accident" governing the shape of the fragments (variations), although he did not mention this particular disagreement with Darwin in his review.

Darwin apparently agreed with Gray's correction.[32] In a letter to Gray shortly after the review appeared Darwin wrote to him:

> You give a good slap at my concluding metaphor: undoubtedly I ought to have brought in & contrasted natural and artificial selection; but it seemed so obvious to me that nat. selection depended on contingencies even more complex than those which must have determined the shape of each fragment at the base of my precipice.—What I wanted to show was that in reference to preordainment whatever holds good in the formation of a pouter pigeon holds good in the formation of natural species of Pigeon. I cannot see that this is false. If the right variations occurred & no others natural selection wd. be superfluous (*CCD* vol. 16, 479 [May 8, 1868]).

Darwin was here referring to the point he had made earlier in the *Variation*:

> If we assume that each particular variation was from the beginning of all time preordained, the plasticity of organisation which leads to many injurious

> deviations of structure, as well as that redundant power of reproduction [viz., superfecundity] which inevitably leads to a struggle for existence and, as a consequence, to the natural selection or survival of the fittest, must appear to us as superfluous laws of nature (vol. 2, 427).

In both of these passages Darwin was thinking mainly about the contingent nature of variation: some variations will be "right," that is, valuable for survival to organisms possessing them, others will be unfavorable, and still others will be of "neutral" value—neither helping nor hindering.[33] Favorable variations will be selected for survival, unfavorable ones will be destroyed, and those neutral in fitness value will be overlooked (J. G. Lennox, 1993, 409–21). If all variations were in the "right direction," in other words "preordained," evolution would not need an assist from natural selection; new varieties would by definition be well adapted to conditions and so would survive. The emphasis in Darwin's mind here is on the randomness of variation with respect to future adaptive needs. This is again an explicit attempt to drive design out of nature's operations. The argument is, if you accept design at the stage of variation, you render superfluous all other natural processes and phenomena—the plasticity of organization, the "redundant power of reproduction," and above all natural selection itself.

Gray rightly disagreed. Darwin surely had committed a logical (if not an empirical) error here. Even if all variations are preordained and designed, Gray would reply, this fact by itself would not necessarily rule out a role for other natural processes and laws, particularly natural selection:

> [The posited laws of nature are] surely not superfluous if "survival of the fittest," "excellent co-ordination," and all the harmonious adaptation and diversity we behold are to result from those very laws.... [This is] the very same difficulty, indeed, and the same impossibility [like the insoluble dilemma between free will and predestination] as that of drawing the limits between the fixed and the contingent, either in the material or the moral world, in which both volition and established order play their mingled parts (*Nation* vol. 6, 236, a passage scored by Darwin in his annotated copy of Gray's review: *CCD* vol. 16, 479–80, n. 5).

Why, Gray was asking, should designed variations rule out other operations in nature governed by natural laws? Darwin, in claiming that they did, did not see the logical possibility of an intelligent designer preordaining variations that he (or it) nevertheless knew were preordained to obliteration by the action of natural selection. That may be a difficult intelligent designer to understand—why would an intelligent designer *do* that?—but the conception is not internally riven by contradiction. Gray was correct on that score.[34]

Gray also answered yet again a question about design that had always puzzled Darwin, as much as any other: how can the universe both be governed by fixed laws and yet still have room for contingency (or "volition," as it is also called)?[35] Gray affirmed his belief that these two seemingly opposite factors "play their mingled parts" in nature. Darwin, perceiving contradiction, could not go along. As he replied to Gray's letter, if stones fallen off a precipice were not specially designed to suit the architect's needs, and if pigeons had not been specially designed to give room for the play of man's fancy, then no more should we believe that any variations in nature are specially designed. Fixed laws explain all, contingency (or volition) nothing.

The last we hear from either Gray or Darwin about the architect metaphor comes in a letter Gray wrote in response to Darwin's May 8, 1869 letter. In it, interestingly, Gray conceded that the "stone-house argument" is "unanswerable in substance," by which he presumably meant that, given Darwin's assumptions, the metaphor was perfectly apt. In this letter he also acknowledged that scientific demonstration was simply not available for showing "design" in variation. This issue was now, as it had always been, the chief bone of contention between the two men. As Gray put it, "the notion of design [in variation] must after all rest mostly on faith, and on accumulation of adaptations, &c" (*CCD* vol. 16, 537 [May 25, 1868]).[36] Gray in fact more or less let Darwin have his way at this point (although he continued his defense of design for another decade in published reviews and other writings). The "stone-house argument" was unassailable, Gray wrote. "So all I could do was to find a vulnerable spot in the shaping of it, fire my little shot, and run away in the smoke" (*CCD* vol. 16, 537 [May 25, 1868]). This is not a memo written by the victor in a sustained battle of words! It is, rather, a parting volley that one expects to miss its mark, as it did with Darwin. But it is also not a surrender: the architecture metaphor has a "vulnerable spot," Gray believed, and as long as one is convinced that the beautiful organisms surrounding us in nature are designed, as Gray did (but never Darwin), the vulnerability will be exposed and exploited.

NOTES

1. The religious controversies surrounding the *Origin* in the years immediately following publication are surveyed by J. Moore (1979), especially 253 ff. Moore takes up Darwin's architect metaphor in the context of these controversies at 275 ff. D. L. Hull (1973) has compiled a selection of scientific reviews of the *Origin* that appeared in the years immediately following publication. J. Lennox (2004) situates the reviews within a larger scientific and theological context.

2. The "metaphor" is often rendered in the literature on Darwin as the "stone-house metaphor," but Darwin preferred to call it an architect metaphor. It may be questioned whether it is

really a metaphor at all; perhaps "analogy" would be a better term. But "metaphor" is the term Darwin used, and the architect is the main character in his use of it.

3. Whether the metaphor was original with Darwin or borrowed by him is not clear. If he did borrow it, possible source are: (a) John Ray's *The Wisdom of God Manifested in the Works of Creation* (1692), read by Darwin in 1838: "[if] a curious Edifice or machine, counsel, design, and direction [resulting in] an end appearing in the whole frame and the several pieces of it, do necessarily infer the being and operation of some intelligent Architect or Engineer, why shall not [we find the same] in the Works of nature [and] infer the existence and efficiency of an Omnipotent and All-wise Creator" (quoted in K. Thomson, *Before Darwin* [2005, 74]). Or (b) Darwin's grandfather, Erasmus, who wrote in *Zoonomia*: "What a magnificent idea of the infinite power of THE GREAT ARCHITECT! THE CAUSE OF CAUSES! PARENT OF PARENTS! ENS ENTIUM!" (1801, 2, 247). Or (c) David Hume: "Stone, and mortar, and wood, without an architect, never erect a house. [The human mind] forms the plan of a house. The original principle of order is [in] the mind, not in matter" (1779, Part II). Word for word, Hume's idea comes closest to Darwin's employment of it.

4. Surprisingly, little systematic attention has been paid in the scholarship on Darwin to his architect metaphor as far as I can tell, although the metaphor itself has been often noted: S. J. Gould (2002, 340–1); J. Browne (2003, 292–3); M. Ruse (2003, 148–9); Desmond and Moore (1991, 543–4); J. Moore (1979, 275). Other discussions of Darwin's metaphors are R. M. Young (1985); Al-Zahrani (2008).

5. The metaphor does work better when the architect is compared to a breeder—a person who does select "intelligently" in choosing domesticated variations for preservation or destruction. Sometimes Darwin did have breeders in mind. For example, the metaphor made its first appearance in a major publication in 1868, in *Variation of Animals and Plants under Domestication*, and the architect here can stand either for breeders or for nature (Darwin recruited the metaphor for both cases). But he appears to have first devised the metaphor (in 1860) when thinking only about natural selection—a non-intelligent process.

6. The Bible, Thomas Aquinas, and John Calvin all employ the image of "God as architect" (in translation of course!). It seems a fairly durable linguistic convention, but it is doubtful that Darwin was consciously drawing upon it from these religious sources. Somewhat oddly, the metaphor was not employed by Paley in *Natural Theology* where it might have been expected. Paley instead selected the words "designer," "contriver," and "Author of Nature." Cf. Erik Høg (2004); and Stephen A. Richards (2006) for additional discussion.

7. That chance might be thought to be the source of variation was a common worry in the 1860s and 70s: "To gaze on such a universe [as ours], to feel our hearts exult within us the fullness of existence, and to offer in explanation of such beneficent provision no other word but *Chance*, seems as unthankful and iniquitous as it seems absurd.... The hypothesis of Chance is inadmissible" (from *Contemporary Review* of September 1875, submitted by "P.C.W" and quoted in Asa Gray *Darwiniana,* 297–8; cf. John Beatty, 2006, 1–14 who analyzes the role of "chance" in Darwin's *Orchids*). For typical statements at the time about "chance" as the only alternative to design, cf. Paley's *Natural Theology*, 40–1, 281–7; Asa Gray's *Darwiniana*, 117, 125–26, 298. Gray represents his own views and those of many other natural philosophers (cf. *CCD* vol. 8, 496 from Darwin to A. Gray [November 1860]; vol. 10, 428 from Gray to Darwin, [August 1862]; vol. 11, 525, from Gray to Darwin [July 1863]).

8. "Chance" (or "accident") has different meanings in Darwin's theory (cf. Lennox, 2004, for a partial survey of usages; and chapter 1 of this volume). In the architect metaphor it refers to one (among several) cause of organic variation, in particular variations that are "unguided with respect to future adaptive needs." Reviews of the issue of chance in Darwin's theory are J. Beatty (1984, 183–211); G. Gigerenzer et al. (1990, 123–162); J. Lennox (2004); S. J. Gould (2002); Millstein (2011).

9. For examples of this criticism from correspondents and Darwin's attempts to meet it, cf. *CCD*, vol. 8, xx, 496; vol. 9, 9; vol. 10, 428; vol. 11, 525; vol. 16, 450–1, 454 and n. 7, and 565–6; and *Collected Papers* vol. 2, 130–2.

10. Asa Gray may have had different reasons than Lyell for objecting to "undirected variation." Lyell refers to "creational laws" as playing a necessary role in evolution, whereas Gray may have been more agnostic about the need for a Creator. But, like Lyell, Gray did want an account of transmutation that could be reconciled with "natural theology," which, however interpreted, amounts to his wish to exclude "chance." Lyell cited Gray with favor in the closing pages of *Antiquity of Man* (1863), as though to suggest that he shared Gray's views. Nuanced differences between Lyell and Gray, and between these two and Darwin, are examined by T. R. Hunter (2011).

11. This particular statement appeared unchanged through all six editions of the *Origin*.

12. A. R. Wallace appears to have had the same hope: "[Occasional] expressions [in the *Origin*] give your opponents the advantage of assuming that *favorable variations* are *rare accidents*... I think it would be better to do away with all such qualifying expressions [such as "accidental" or "rare"], and constantly maintain [that] *variations* of *every kind* are *always occurring* in *every part* of *every species*" (*CCD* vol. 14, 229 [July 2, 1866], original emphasis).

13. The *Variation* is actually only a part of the planned "big species" book. The rest of it appeared in published form only in 1975, under the title *Natural Selection,* under the editorship of R. Stauffer.

14. The seeds for Darwin's architect metaphor, like so many other ideas in his mature theory, may have been planted much earlier than his first mentions of it in his correspondence in 1861. He records in the "Old and Useless Notes" (1838; hereafter "OUN"), just after he read Mayo's *The Philosophy of Living* (1838) that he liked very much Mayo's statement that "architecture is a fine amplification of two ideas in nature: a developement [*sic*] of the thoughts expressed in Fingals cave, & in the arched and leafy forest" ("OUN" 6). The first of these two ideas may well correspond to a "thinking" architect, the second to "naturally produced materials" for the architect to work upon. Just a few pages later in the same "OUN" (11v) Darwin states: "[Reynolds] says architecture does not come under imitative art; (my view says yes—mass of rock)." A "mass of rock" is just what Darwin later (in the metaphor) likened to "accidental or random variations" in living creatures. See n. 3 above.

15. W. H. Harvey, in commenting on the *Origin* in 1860, employed the very same metaphor, apparently without any knowledge that Darwin had used it himself: "I am willing to admit that [your theory] explains several facts which are not otherwise easily to be accounted for. [But] I can only regard Natural Selection as one Agent [in biological evolution] out of several;—a handmaid or wetnurse—so to say—but neither the *housekeeper*, nor the mistress of the house" (*CCD* vol. 8, 322 [August 24, 1860], emphasis in the original). The image of a "handmaid" to a master or mistress was earlier employed by Thomas Hobbes in 1651 to describe the relation of "reason" (the handmaid) to "appetite" (the ruling power in the human soul).

16. Lyell persisted in this sentiment in his *Antiquity of Man*, 1863, using nearly the same expression: "In our attempts to account for the origin of species we find ourselves still sooner brought face to face with the working of a law of development of so high an order as to stand nearly in the same relation as the Deity himself to man's finite understanding, a law capable of adding new and powerful causes such as the moral and intellectual faculties of the human race to a system of nature which had gone on for millions of years without the intervention of any analogous cause. If we confound Variation or Natural Selection with such creational laws we deify secondary causes or immeasurably exaggerate their influence" (*AM*, 1863, 469).

17. *CCD* vol. 10, 438–9 (August 1862) to Lyell: "Falconer does not understand me that NS can do nothing without var, governed by the most complex laws." (Cf. vol. 10, 440 to Falconer, where Darwin insists on prior variation for natural selection to work, governed by many "unknown laws"; e.g., elephants do not get their trunks from direct action of external conditions but from "some unknown law").

18. Hooker had already detected Harvey's confusion as early as May 1860, when he pointed out to Darwin that Harvey, in a privately printed pamphlet, had suggested that Harvey seemed to think Natural Selection is somehow logically dependent on variation (*CCD* vol. 8, 249 [June 8, 1860].

19. *CCD* vol. 8, 258 [June 17, 1860], to Lyell: "It is only owing to Paley & Co., as I believe, that [finding special divine interference] is thought necessary with living bodies.—But we shall never agree, so do not trouble yourself to answer."

20. Lyell had earlier objected to a passage in Huxley's review of the *Origin* in the *Westminster Review* in which Huxley claimed: "Matter and Force are the two names of the one artist who fashions the living as well as the lifeless" (in *CCD* vol. 8, 260). Lyell objected to the omission of any First Cause in this conception.

21. For references to Darwin's conclusions about free will in the *Notebooks*, cf. *inter alia CDN* M 27, M 31 (which contain the intriguing statement that "free will and chance are synonymous"), M 72 (in which Darwin discussed the "free will of an oyster"), and *Notes* 25 ("Free will [in man] is not present"). Additional discussion follows in chapter 9 of the present work.

22. For bibliographic details of Darwin's *Variation*, cf. *CCD* vol. 11, 583, n. 13.

23. Darwin apparently attached some weight to the consideration that a divine designer would never have created varieties just to suit man's fancy, and so any supposition that He created the whole range of varieties that we find in nature fails by the same criterion: cf. *CCD* vol. 8,161 [April 1860] to Lyell; vol. 9, 162 [June 5, 1861] to Asa Gray; vol. 9, 234 [July 1861] to Lyell; vol. 16, 479 [May 8, 1868] to Asa Gray; and *Variation* vol. 2, 427.

24. The editors of the *CCD* surmise that Darwin selected the example of pigeons because in them "accidental variation had been 'extraordinarily great'" (vol. 11, 583, n. 11; cf. *Variation* vol. 1, 223, where Darwin refers to "sudden and apparently spontaneous variation"). Pigeons are now known to vindicate Darwin's suspicions that they exhibit "extraordinarily great" variations, compared with other species. Cf. Michael Shapiro, et al., *Science DOI: 10.1126/science.1230422*, 2013.

25. Excerpts from Matthew's contribution to the species question (as reported by him in *Gardener's Chronicle* in April 1860), Darwin's response, and Darwin's correspondence related to the issue are contained in *CCD* vol. 8. In a letter to Matthew in June 1862 Darwin refers to him with the grand compliment as "the first enunciator of the theory of Natural Selection"

(*CCD* vol. 10, 251 [June 13, 1862]). For discussions, see Mayr (1982), Eiseley (1961), Dempster (1983, 2005), and Gould (2002).

26. Stones, of course, do not die; organic beings do. The metaphor, thus, fails on this score as well. Natural selection *can* be said to "destroy" as well as "preserve" and even "fashion" new species. The most an architect can do is to *ignore* fragments that do not suit his purposes, but the ignored stones continue indefinitely to exist. Darwin never did acknowledge this defect in the metaphor, and if he did notice it, he did not think it fatal to its illuminative power.

27. The passage is often quoted in the secondary literature, usually in support of the argument that Darwin placed supreme importance on natural selection at the expense of other causal factors in evolution, particularly variation. I claim the passage in larger context does a good job of highlighting just how important Darwin believed variation to be, how he conceived variation to act in concert with selection, and how he wished to cast its role in his theory as no more worrisome for a theological critic than stones that fall off cliffs.

28. A parallel change in wording was making its appearance at about the same time in the *Origin*. "Spontaneous variation" began replacing the expressions "accidental variation" and "chance variation" beginning with the fourth edition of the *Origin* (1866). By the sixth edition (1872) "spontaneous variation" assumes almost the entire weight of what had previously been the quite common "chance variation."

29. See chapter 1 of this volume on the distinction between "chance" as "probability" and "chance" as "random with respect to future needs or purposes." Here Darwin has the second usage in mind.

30. Gray's response to the architect metaphor may have been friendlier than responses of Darwin's British sympathizers (like Lyell) because of Gray's Calvinism. As has been shown by James Moore (1979), the more orthodox the religious beliefs of particular readers of *Origin*, paradoxically, the more likely they would be to give Darwin's theory a favorable reception.

31. Another metaphor employed by Darwin: in 1862 Darwin complimented Gray for his review of his *Orchids* book by saying "Of all the carpenters for knocking the right nail on the head, you are the very best" (*CCD* vol. 10, 330–1 [July 23, 1862]). This was in reference to Gray's having perceived that *Orchids*, with its reliance on the notion of "contrivances," was a "'flank movement' on the enemy [viz., the intelligent designers]." Cf. J. Beatty, 2006, 1–14.

32. Darwin's agreement with Gray's "slap" at his metaphor was given only in private correspondence. Darwin retained the image unaltered in the second edition of *Variation* (1875).

33. The importance of "neutral" variations in evolution has been brought to the forefront of recent discussions, thanks largely to the work of M. Kimura (1981, 1982) and his followers. Cf. G. Gigerenzer (1989, 158) for discussion.

34. Another reviewer of the *Variation* in an Edinburgh newspaper saw one possible way to render the possibility that Gray was suggesting: "We doubt not that Professor Asa Gray... could show that natural selection, supposing it proved as operative, is simply an instrument in the hands of an omnipotent and omniscient Creator" (quoted in *CCD* vol. 16, 480, n. 6). I do not believe this is the route Gray would have wanted to take. He accepted "blindness" for the process of selection, insisting on intelligence only for the prior production of variations. If one must have an omniscient Creator, and seemingly Gray did, its role in evolution would be not to "select" favorable variations but only to design them and "foresee" those variations that would be selected by "blind" processes.

35. The idea in Gray's letter that volition is equivalent in some way to contingency had been expressed many years earlier in Darwin's private notebooks: (*CDN* M 31). Gray certainly had no direct knowledge of this or other notebook passages, so the coincidence in usage and conception is striking.

36. Gray had already made it clear as early as October 1860 that he saw only two alternatives for the causes of variation: design or chance: "So the issue between the skeptic and the theist is only . . . whether organic Nature is a result of design or chance" (*Darwiniana*, 125). Gray's opinion on this is given on the next page: "a fortuitous Cosmos is simply inconceivable." Gray continued to think that the "chief difference" between his views and Darwin's centered on the role of "chance" in variation. Gray could not go along with chance at all, and Darwin could not go along with design. Darwin always believed variations were due to the operation of natural laws. He often waffled on the question of whether those laws were themselves divinely ordained. Cf. *CCD* vol. 11, 525 (July 7, 1863), letter from Gray to Darwin; vol. 8, 274, letter from Darwin to Asa Gray (July 1860).

8

Darwin's Giraffes

INTRODUCTION

When Darwin published the first edition of his masterpiece *Origin of Species* in 1859 he barely mentioned giraffes. They appear twice, only to say that their tails are well suited to swatting flies and that they have the same number of vertebrae as elephants (*Origin, Variorum* edition, hereafter *Variorum*, 322: 9; 362: 184; 745: 174; the example also appears in *Charles Darwin's Natural Selection*, 378). By the sixth edition of the *Origin*, published in 1872, this had changed. Now giraffes figure prominently, especially in the newly added Chapter VII in which they appear over a dozen times.

Darwin wished to show that the giraffe's striking features, especially its long neck, far from challenging his theory, as had been suggested by a hostile critic,[1] actually gave it additional support.

How Darwin turned a hostile example into a friendly one is interesting in itself. But the additional paragraphs also give an important illustration of how Darwin's thought and expression about the origin of new species had evolved from the first edition. In particular, light may be shed on whether his thought had changed significantly over the years and whether it changed in a "Lamarckian" direction, as many scholars believe.[2] That question in turn helps answer others, such as which edition of the *Origin* should be regarded as the canonical one[3] and what Darwin himself thought "Darwinism" in its mature form should mean. Darwin's carefully crafted sentences about giraffes are an opportunity for giving some important clarifications.[4]

Giraffes entered nineteenth-century Europe through two routes—one physical, the other conceptual. The physical route was that taken by the prominent French zoologist Étienne Geoffroy St. Hilaire, who personally accompanied the first living giraffe to come to France (a gift to France from the Pasha of Egypt) on a journey across the country from Marseilles to Paris in 1826.[5] The conceptual point of entry was J. B. Lamarck who, in his *Philosophie Zoologique* (1809, translated into English as *Zoological Philosophy*, hereafter *ZP*), employed giraffes and their long necks to provide support for his theory of transmutation.[6] Lamarck—much to his later regret—became notorious for this example, but by it he entered into the popular imagination and his name has been closely linked with giraffes ever since. Geoffroy's giraffe, by contrast, was soon forgotten. It is likely by way of Lamarck that Darwin first came to think about giraffes in an evolutionary sense[7] (he mentioned them briefly in *Journal of Researches*, but only to say they had been observed in Southern Africa by Andrew Smith),[8] and no doubt by way of the same source that his hostile critic, St. George Mivart, came to think about them as well. Thus, any account of Darwin's giraffes necessarily implicates Lamarck whether Darwin wished it to be so or not.[9]

That Lamarck's views are often associated with giraffes is perhaps unfortunate.[10] His theory of evolution is more complex than what the giraffe example suggests about it, and in any case the neck of the giraffe was intended to be only an illustration, not the most important one and certainly not the iconic case that it has become.[11] If one wanted examples one would do better to look at what Lamarck had to say about the adaptive features of the feet of birds. Nevertheless, I shall stick with giraffes because: (1) "Lamarckism" is today, for better or worse—(cf. Burkhardt [1977], 201)—often associated with them, as Gould admits when he calls the neck of the giraffe "our canonical just-so story" ("The Tallest Tale," *Natural History*

[1996], **105**, 22); 2) Darwin paid especially close attention to them in the *Origin* sixth edition; and (3) Darwin did so because he believed the giraffe's long neck illuminated important points about variation and natural selection, in other words about "Darwinism," and so, by implication, about how "Darwinism" differs from "Lamarckism."[12]

DARWIN AND LAMARCK

In late 1837 or early 1838 Darwin commented privately to himself in his "B" *Notebook*, "My theory very distinct from Lamarcks" (*CDN* B-214).[13] This was well before he read Malthus (September and October 1838), an encounter he reported many years later as having been transformative in his thinking about the species question (*Autobiography*, 1876 [1958], 120; cf. Kohn, 1980; La Vergata, 1985, in Kohn, ed., 1985; S. Herbert, 1971). Darwin, apparently, did not need Malthus to see that he had "a theory" of his own (many references in 1837–1838 to "my theory" in the *Notebooks*).[14] Nor did he need Malthus to see that his theory was distinct in important ways from the theory of Lamarck.[15] (Darwin did not see, even privately, a need to establish his originality against either Erasmus Darwin or Robert Chambers; when he compared his theory to someone else's, it was almost always Lamarck's.) What did Darwin believe the distinction was?

His answers were mostly hidden from public view by being inserted into marginal comments in works he read or, somewhat more publicly, private letters of correspondence. From the evidence of these works—works that we should note Darwin did not anticipate ever being published—it is clear that Darwin saw more than just one difference between his theory and that of Lamarck.

One important difference, pushed home with special vigor in his correspondence with Charles Lyell, is that Darwin from an early time was no "progressionist," as Lamarck was. That is, he did not believe that organisms, as they descended with modification from earlier progenitors, made any sort of necessary "progress" up a "ladder of evolution." Darwin saw a "tree" or "coral" of life, not a ladder, and this view allowed him to think of species as moving "forward," "backward," or even "sideward"—whatever works to secure survival in a difficult and ever-changing environment. Indeed, one of the "revolutions" wrought by Darwin was to problematize the very notion of "progress" as applied to organic productions (e.g., *CDN* E-95).[16]

But Darwin saw another difference from Lamarck, one that he was more reluctant to speak about openly but that more clearly establishes his theory as the one that would triumph. Briefly, for Lamarck, organisms evolved by means of two mechanisms: one, an innate "progressive" impulse or instinct moving all living forms upward along an ascending scale of perfection whose most recent (but not necessarily final) instantiation is "man;" and, secondly, a capacity in organisms to adapt to local conditions as these change, so as to ensure survival in a changing environment.[17] What is important here is that Lamarck did not require "chance" in the operation of his theory on either mechanism, and even excluded any possible role for it. If organisms have an innate tendency to progress, chance is not needed to explain their progression. Their progression is explained by an "innate tendency," and the tendency, being innate, ensures progression. And if organisms have a capacity to "adapt" to changing conditions, no chance is possible here either. Rather, adaptation is a straightforward process of organisms encountering new environments and making necessary—even predictable—adjustments to their habits and structures (Lennox, 2004).[18]

For Darwin, evolution did not work like this. He disallowed progression, and he had an entirely different explanation for successful adaptation. As to the former, organisms change in various directions, not just "upward," whatever that expression might mean. Indeed, Darwin's theory throws into question directionality of change of any sort except "good" and "bad" directions. "Good" are those that assist in an organism's survival, "bad" are those that work against survival. The key question here is, will a variation succeed over its rival forms in the struggle for survival? Sometimes more primitive, or less complex structures will do better than more "advanced" and complex forms.[19]

As to adaptations, Darwin's thought was more complex and subtle than Lamarck's. Creatures for Darwin do not simply "make adaptations" to their environments. Rather, they either happen or happen not to be well suited to environmental conditions. If they chance to be well suited, they are more likely to survive than if they are not well suited. In the context of this question Darwin employed "chance" in different ways. One may say, as Darwin often did, that some organisms have a "better chance" of surviving than others. Chance here means only "probability" or "likelihood." As indicated above, it is a statistical, not a stochastic concept.

Darwin knew of Lamarck's transmutationist theory both from indirect testimony from some of his mentors and, in the 1820s and 30s, from reading Lamarck himself. It is often forgotten that he had some admiration for the French zoologist, occasionally giving him much credit for original ideas and observations.[20] Darwin even used many of Lamarck's examples to illustrate his own theory of transmutation—the reduced wings and enlarged legs of domesticated ducks due to use/disuse, the aborted eyes of burrowing creatures, the aborted wings of some insects, and others.[21]

Nevertheless, Darwin was certain from an early age that his own "theory" was different from Lamarck's, and he rated Lamarck's views in the *Philosophie Zoologique* as "poor" and "useless."[22] It is the latter view that tends to be remembered, thanks mainly to Darwin's often contemptuous remarks.[23] But, while his own theory was indeed quite different from Lamarck's in important ways, Darwin did draw on Lamarck for significant insights. From his earliest musings on origins to his last he remained committed to the Lamarckian elements of "direct effect of physical conditions on habit, and thereby on organization," "use and disuse," and "inheritance of acquired characters" in effecting organic change. These three elements of Lamarck's theory, sometimes referred to in shorthand as "use-inheritance," are not the whole of what might be called "Lamarckism." But these are the ideas that Lamarck is remembered for today, so much so that the expression "Lamarckism" in conventional usage is essentially encapsulated in these three factors.[24]

Darwin accepted use-inheritance in all editions of the *Origin* and in his 1868 work *Variation of Animals and Plants under Domestication*.[25] His favorite example to illustrate it was what he believed happened to ducks as they became domesticated. They were forced to use their legs more (i.e., "use") and their wings less ("disuse"), and this, over time, produced a gradual increase in the size of leg bones and a decrease in the size of wing bones.[26] Such changes, Darwin believed, were passed down to offspring. He gave many other examples in the *Origin*. If one believes that the earth's environment changes over time—and Darwin, as a convinced Lyellian, did believe this—then one may also believe in environmentally induced heritable organic change, and Darwin often did.

Darwin did not, however, believe this always happened. For example, the larger arm muscles of the iron-smith brought about by the practice of the trade are not, Darwin thought, transmitted (*Variation*, vol. 2, 287). For a modification of organization to be passed down it must be acted upon by changed conditions over many generations. But even then heritability may fail. For instance, certain breeds of dog that have had their tails cropped generation after generation nevertheless continue to produce offspring with tails.[27] The practice among some peoples of circumcision is another example mentioned by Darwin.[28] In general, Darwin did not accept the heritability of what he called "mutilations" (cf. *Natural Selection* 294; *CCD*, vol. 17, 59 [February 1869] to W. C. Tait; vol. 17, 225 [April 1869] to Linnaean Society; vol. 17, 229 [May 1869] to J. J. Weir: "I am hardly a believer in inherited mutilations [as cropped dogs' tails]"). On the other hand, some acquired characteristics are "without doubt" transmitted: the drooping ears of dogs from disuse, the modified intestines of some farm animals from "use" (*Variation*, vol. 2, 83, 290, 346, 367), and, as always, the enlarged legs of domesticated ducks from disuse of their wings. These passages show that into late life Darwin retained Lamarckian elements in his own theory.

But did Darwin become *increasingly* Lamarckian as the years wore on, and if so, did he come to alter his own theory to make accommodation?[29] Some of his private statements to correspondents suggest that he thought he did, while others suggest otherwise.[30] At times Darwin was explicit that he did not see much change. In the final edition of the *Origin* that he edited for publication (1872) he wrote, "I have *always* maintained [the Lamarckian factors of use and disuse] to be highly important" (*Variorum*, 242: VII.382.65.0.50.2–6, emphasis supplied).[31] Moreover, the role of natural selection in evolution, in comparison with use-inheritance, always remained paramount, and was arguably even strengthened in later editions.[32] The best bet is probably that Darwin was not sure about this himself. His memory could, and sometimes did, play tricks on him, and in any case the question is difficult. His discussions of giraffes help shed some light.

GIRAFFES IN LAMARCK AND THE *ORIGIN*

Lamarck was notorious for his opinion that giraffes have long necks because they make a deliberate effort to reach higher.[33] Through use (i.e., habit) over many generations, they acquire longer necks, and these are then passed down to offspring. Organisms, usually the young, change through a willed effort, necessitated by changed physical conditions. The "effort" is prompted by an interior sentiment (*sentiment interieur*) that translates first into changed habits and thereby into a change of structure and organization.[34] In the case of ruminants, including giraffes, the change in conditions seems to have come about by the appearance of new predators, including "man." This in turn gave rise to "fits of anger," leading to "efforts of their inner feelings [to] cause the fluids to flow more strongly towards that part of the head [to produce] bony and horny matter" (*ZP*, 122). These new organs were then used to fight off predators. The same reasoning applies to the question of how the giraffe attained its great height:

> It is interesting to observe the result of habit in the peculiar shape and size of the giraffe: this animal lives... where the soil is nearly always arid and barren, so it is obliged to browse on the leaves of trees and to make constant efforts to reach them. From this habit long maintained in all its race, it has resulted that the animals' forelegs have become longer than its hind legs, and that its neck is lengthened to such a degree that the giraffe... attains a height of six metres (*ZP*, 122).

The important thing here is the *cause* of change in structure: a change of habit that is induced by changes in the environment. Obviously an organism has to be alive before it can alter its habits.[35] Why it alters its habits has to do with fluids, secretions, and sentiments, or feelings. The organism is "obliged" to change its habits and it must make "constant efforts" to change. The mechanism of change is the effort (not necessarily conscious!) of the organism to meet the demands of the environment. Lamarck then goes on to say that the longer necks thus acquired are passed down (*ZP* 1984 [1809], 107–8).

What did Darwin think about this account? He did not discuss giraffes in detail in any edition of the *Origin* prior to the sixth. He did note (in the third edition) that Lamarck had attributed the long necks "to use and disuse, that is to habit" (*Variorum*, 60: 12), but this observation, which appeared in the "Historical Sketch" (retained unaltered after it first appeared in the third English edition), is made without editorial comment. Darwin neither agreed nor disagreed with Lamarck, only reported what he thought Lamarck believed, in keeping with the overall strategy of the Sketch to give a brief "history of opinion" on the species question. He also, in the first edition, drew attention to some peculiar adaptations and other interesting features of the giraffe, as noted above. But these passages, retained through every edition, do not address the Lamarckian question of the *causes* of these features. They are simply given as features that are interesting and noteworthy.

In the sixth edition, however, Darwin addressed the Lamarckian issue head-on. Giraffes come up more than a dozen times, a feature that is new to the sixth edition, with all instances (not surprisingly) in the new Chapter VII, and in direct response to highly critical reviews of the *Origin* published by St. George Mivart in 1869 and 1870 in the *Quarterly Review* ("a most cutting Review of me") and the *Month*.[36] Mivart had challenged Darwin's belief that "chance" plus natural selection could account for much if any change:

> [It is to be found] each successive year that deeper consideration and more careful examination have more than brought home…the inadequacy of Mr. Darwin's theory to account for the preservation and intensification of incipient, specific, and generic characters. That minute, *fortuitous,* and indefinite variations could have brought about such special forms and modifications as have been enumerated…seems to contradict not imagination, but reason (Mivart, [1871], p. 74, emphasis added; cf. pp. 90, 93, 125 for similar expressions; and *CCD* vol. 19, 111 and n. 6 for Darwin's reply to Mivart in a letter to his son Francis in 1871).

Mivart's star witness for his claim was, somewhat surprisingly, the giraffe. He found several flaws in Darwin's arguments for natural selection, and concluded that something else besides natural selection must be going on in nature (*Genesis of Species* [1871], 36–41). For Mivart this "something else" was, to put it briefly, an "unknown natural law" that ensures harmonious and coordinated variations in the right direction at the right time for evolution to occur, thus implying an important role for a Designer.

Mivart does not say what drew him to giraffes in his critique of Darwinism. Other naturalists besides Lamarck had written about giraffes, most notably Richard Owen, the only one as far as I can tell who actually connected them to Darwin's evolutionary hypothesis prior to Mivart.[37] Owen commented that more evidence from the fossil record is required to believe that the long neck of the giraffe evolved from earlier forms (*Paleontology* [1860], 406). Darwin read Owen's *Paleontology*, both editions, in 1862 (*CCD* vol. 10, 16–17 and nn. 5–6; also xxvi) but did not score this passage. Perhaps Mivart read it too—Owen did anticipate Mivart to some extent—or he may have read the anonymous review of Owen's book in the Catholic periodical *The Rambler* (volumes 3–4, 1861), 127 (Mivart was Catholic). The reviewer drew attention to Owen's comments, noting that Darwin's theory could not account for features like the giraffe's long neck—essentially the point Owen had hinted at and that Mivart developed. Darwin read Mivart's 1871 book almost as soon as it appeared in published form (*Marginalia* 584–9), and felt the sting of Mivart's criticism of how he had handled giraffes, for he immediately sent letters to Mivart challenging his interpretation. (The letters exchanged between the two men on this subject are printed in *CCD* vol. 19, various dates.)

From the letters, and others that Darwin sent to Hooker, Huxley, and his son Francis, one sees quite clearly that Darwin believed Mivart was both dishonest and unfair about his theory in general and about his treatment of giraffes in particular (e.g., *CCD* vol. 19, 578–9 and n. 4). He even went so far as to pay for the private publication of a pamphlet written by the American mathematician Chauncey Wright in which Wright defended Darwin against Mivart, quite ably in Darwin's opinion (*CCD* vol. 19, 453). Darwin's "bulldog," T. H. Huxley, also came to Darwin's defense against Mivart in the *Westminster Review* in 1871. Darwin was grateful for these interventions. A full examination of this unfortunate episode would require a separate chapter. Suffice it to say that Darwin felt betrayed by Mivart. The man had shown exceptional cordiality to Darwin in personal meetings ("we were the best of friends [in personal encounters]," remarked Darwin [*CCD* vol. 19, 110–1]), but attacked him with viciousness in published works (often anonymously: cf. *CCD* vol. 19 [1871] 609–10, letter to CD from Hooker). Darwin had some difficulty letting go of his negative feelings toward Mivart; no one had quite gotten under Darwin's skin as much as he had (*CCD* vol. 19 [1871], 591).

Darwin was disturbed by several aspects of Mivart's criticisms of his theory: one, that he had been misquoted; two, that Mivart took important statements out of context; three, that Mivart had failed to notice that Darwin did not assign *everything* to natural selection (here Darwin insisted that he had always given some room to Lamarckian use-inheritance; e.g., *CCD* vol. 19, 31–2 and nn. 3 and 7, in a letter to Francis Darwin; and ibid., 34, in a letter to Mivart; ibid., 50, in a letter from Darwin to A.R. Wallace; and ibid., 478 in a letter from Darwin to A.R. Wallace. Wallace responded to Darwin, with specific reference to giraffes, in *CCD* vol. 19, 482–3 and n. 3).

But Darwin also perceived, correctly, that Mivart was attacking his theory at the point of "fortuity" in variation. Lamarck's theory at least had the advantage of ruling out fortuity, even if not explicitly ruling in design. "Use/disuse" suggests a role for calculated or "willed" adaptation, and there is nothing fortuitous about this. The question Darwin considered is, did giraffes come to have their long necks through increased use to reach higher branches and then pass this modification down to offspring? That would be a Lamarckian explanation. Did Darwin accept it?

Several of the passages about giraffes added to the *Origin* do not shed much light. Often Darwin chose only to describe the giraffe's "beautiful adaptations" and how these serve its survival needs (e.g., *Variorum*, 242: *VII.382.65.50. 11–12*; 243. *VII.382.65.0.50.17*; 243: *VII.382.65.0.50.26–27*). If anything, the implication is against Lamarck in these passages, in the sense that Darwin in the latter two refers to the advantage of longer necks for the "nascent," that is the emergent giraffe. (The term "nascent" had been introduced by Mivart to ridicule the idea that nascent features could be adaptive and so "selected.") Darwin's idea seems to be that the first longer-necked creatures, the ones that will in time become giraffes, are *born* with the advantageous longer necks, not that they *acquire* them after birth:

> So under nature with the nascent giraffe, the individuals which were the highest browsers and were able during dearths to reach even an inch or two above the others, will often have been preserved; for they will have roamed across the whole country in search of food.... Slight proportional differences [in members of a species], due to the laws of growth and variation, are [often] not of the slightest use or importance to most species. But it will have been otherwise with the nascent giraffe, considering its probable habits of life; for those individuals which had some one part or several parts of their bodies rather more elongated than usual will generally have survived (*Variorum*, 243: VII.386.65.0.17–20).

From this (and similar passages) we cannot say with certainty *how* the longer necks arose initially. We are told only that slight individual differences are "due to the laws of growth and variation," not what those laws are.[38] Certainly Darwin did *not* say the longer necks resulted from "increased use," much less "willed effort." It is actually more natural to read him as saying that the longer necks came first, for no assignable reason, leading to new habits useful for survival, rather than the other way round.[39]

However, when we look further into Chapter VII, we may be tempted to find some fudging in a Lamarckian direction:

> I. With the giraffe, the continued preservation of the individuals of some extinct high-reaching ruminant, which had the longest neck, legs, &c., and could browse a little above the average height . . . would have sufficed for the production of this remarkable quadruped; but the prolonged use of all the parts together with inheritance will have aided in an important manner in their coordination (*Variorum*, 262: *VII.382.65.0.50.300*).
>
> II. By this process [of natural selection] long continued . . . combined no doubt in a most important manner with the inherited effects of the increased use of parts, it seems to me almost certain that an ordinary hoofed quadruped might be converted into a giraffe (*Variorum*, 243: *VII. 382.65.0.50.23*).
>
> III. Mr. Mivart then asks [that] if natural selection be so potent, and if high browsing be so great an advantage, why has not any other hoofed quadruped acquired a long neck and lofty stature, besides the giraffe? [To which I respond that] in every district some one kind of animal will almost certainly be able to browse higher than the others; and it is almost equally certain that this one kind alone could have its neck elongated for this purpose, through natural selection and the effects of increased use (*Variorum*, 244: *VII.382.65.0.50.32, 36*).

Here in fact we do get the important word "use," and in the first two passages the idea that modifications acquired by increased use may be inherited. The question is, what is to be made of this? It must be noted, first of all, that in all three passages (as throughout the new Chapter VII), Darwin is engaged in a polemic against Mivart's claim that natural selection cannot possibly be as potent a force in evolution as Darwin asserts for it. In all of these passages, Darwin is mainly saying, "yes, it can be and in fact is." This only underscores what he had said repeatedly in the first edition: "Over all these causes of Change [including "use and disuse" of parts] I am convinced that the accumulative action of Selection . . . is by far the predominate Power" (*Variorum*, 119: 322).[40] In addition, he had already in the first and all subsequent editions allowed *some* role in variation for use-inheritance, so to reassert *some* role for use-inheritance here is neither new nor surprising. It still plays a subsidiary role (even if an "important" one) to that of natural selection. Overall, the passages on giraffes (and other creatures too, for that matter) in Chapter VII say no more than this.

More to the Lamarckian point, however, is what these passages say about the *causes* of variation. Actually, they say very little, and certainly fall short of saying what a Lamarckian must say, namely that changed environments cause changed habits (i.e., "use"), which cause changes in structures, and these are then transmitted to offspring. The only part of this series of causal steps included in any of the above passages is the assertion that the effects of "use" are inherited by offspring. The idea seems to be that modifications of structures come about by "unknown" causes, that these modifications are in some degree "*aided*" by "use," and that the alterations thus aided are then passed down.[41] "Use" by itself in these passages does not give rise to new variations that are heritable, but that is the key point on which Lamarck insists.[42]

This construction of Darwin's meaning may be defended by looking at the first passage quoted above (I). The passage has two parts: first, the appearance (cause not given) of a ruminant that "could browse a little above the average height"; second, the "prolonged use of all the parts together with inheritance." These are different causes, as indicated by Darwin's use of the conjunction "but" to separate the two phenomena. We might well read the passage as saying that the original cause of the longer neck is merely accidental variation (i.e., cause unknown), but that once this variation appears it is "aided" (or augmented) by continued use. Look at the very next sentence:

> With many insects which imitate various objects, there is no improbability in the belief that accidental resemblance to some common object was in each case the foundation for the work of natural selection (*Variorum*, 262: *VII.382.65.0.50.301*).

"Accidental resemblance" explains the initial appearance of some feature, "use" then may augment or assist what accident has produced, and natural selection then works on the modified part in its various stages of development (cf. *Variorum* 243: *VII.382.65.0.50.22–3; and* 261: *VII.382.65.0.50.282*). This gives the best account of how Darwin viewed the process of speciation whenever "use" did play a role.

In fact Darwin had good theoretical reasons for doubting that use was a common cause of variation. "Use" will be acted upon by natural selection only when a variation induced by use is "beneficial" to the organism in question. "Habit [i.e., use]," Darwin remarked, in speaking of the prehensile tail of some monkeys, "almost implies that some benefit great or small is derived [from variations brought about by this cause]" (*Variorum*, 253: *VII.382.65.0.50.172*). It does not make any sense to speak of an organism "using" a particular structure or organ to *decrease* its chances of survival. In other words, "use" implies that an organism is "trying to adapt" to its circumstances, although of course in Darwin's view no "will" or "conscious effort" is involved.[43] Yet no organism would try to adapt in a way that is antithetical to its survival needs. Darwin believed, however, that many variations are *not* useful for survival, and that many are even deleterious.[44] There is no guarantee, he wrote, that variations will always be "in the right direction and to the right degree" (*Variorum*, 245: *VII.382.65.0.50.45*; cf. *CCD* vol. 8, 340 and 355 [September 1 and 12, 1860] to Lyell; and letter to Hooker [September 2, 1860] 342; and *Natural Selection* 234, 252, 262, 283). When variations are not in the right direction they will not be preserved, unless they are neutral in survival value. This plainly suggests that many variations are simply random with respect to adaptive needs, not the effects of use and habit.

THE "ACCIDENTAL" ORIGIN OF GIRAFFES IN THE *ORIGIN*

To this point we have seen that Darwin was willing to allow a role for use in *aiding* variations once they had already appeared—e.g., a ruminant born with a slightly longer neck might, through effort, succeed in stretching the neck a bit more, and then might pass the longer neck down to offspring.[45] That is Lamarckian, but it is by itself not new in the sixth edition; it had been there from the start. What, though, do Darwin's giraffe passages suggest about how the variations arose in the first place? The best inference is that Darwin wished to suggest "chance" (or "so-called spontaneous variation") without actually mentioning this cause by name. Let me provide representative instances:

> I. Why, in other quarters of the world, various animals belonging to this same order [viz., ungulate] have not acquired either an elongated neck or a proboscis, cannot be distinctly answered; but it is as unreasonable to expect a distinct answer to such a question, as why some event in the history of mankind did not occur in one country, whilst it did in another (*Variorum*, 244: *VII.382.65.0.50.38*).
>
> II. Why [S. Africa but not S. America] abounds with [giraffes] we do not know; nor why the later tertiary periods should have been much more favorable for their existence than the present time. Whatever the causes may have been, we can see that certain districts and times would have been much more favorable than others for the development of so large a quadruped as the giraffe (*Variorum*, 244–5: *VII.382.65.0.42–3*).

Directly following these passages on giraffes, and in the same context of wondering why certain variations come about when and where they do, Darwin wrote about insects:

> III. But in all the foregoing cases the insects in their original state no doubt presented some rude and accidental resemblance to an object commonly found in

the stations frequented by them. Nor is this at all improbable, considering the almost infinite number of surrounding objects and the diversity in form and color of the hosts of insects which exist (*Variorum*, 247: *VII.382.65.0.50.74–5*).

IV. Assuming that an insect originally happened to resemble in some degree a dead twig or a decayed leaf, and that it varied slightly in many ways, then all the variations which rendered the insect at all more like any such object, and thus favored its escape, would be preserved, whilst other variations would be neglected and ultimately lost (*Variorum*, 247: *VII.382.65.0.50.77–8*).

Let us take these pairs of passages in reverse order. In passages III and IV, instead of asking why giraffes have long necks, Darwin is asking, why do many insects have excellent camouflage? In other words, where and how do variations favorable to survival originate? The insects passages both invoke the idea of "chance" almost as clearly as could be hoped without using the actual word (which, in the sense of "unknown cause of variation" Darwin had by this time all but expunged from the *Origin*).[46] Passage III makes reference to "some rude and accidental resemblance" of an insect's coloration to its environs, Passage IV to variations that "originally happened to resemble in some degree" surrounding conditions. "Accident" and "happenstance" suggest "chance" as the "cause" of variation. It is not difficult to read this explanation back to Darwin's just completed discussion of giraffes: creatures that "happened to have" longer necks that were heritable would be selected for survival. That, in fact, is the most straightforward way of reading Darwin's meaning throughout the new Chapter VII wherever "causes of variation" is the question.[47]

Passage IV brings to attention another primary concern that Darwin faced in addressing Mivart's criticisms. Recall that Mivart was not so concerned to fault Darwin for failing to be sufficiently Lamarckian as he was to accuse Darwin of overestimating the role of natural selection in evolution. Mivart had claimed that "variability" by itself, if properly directed and unassisted by natural selection, could produce most if not all the change we see in nature.[48] Variability was in Mivart's view decidedly *not* fortuitous, but rather was guided by some unknown innate principle. Darwin employed the insect example to refute Mivart's claim and to reassert the power of natural selection in the process. Variation of course must come first, to supply the "raw materials" on which natural selection can then act, and these variations are "accidental." But variations, even heritable ones, will not usually lead to lasting changes in organisms unless they are selected. The camouflage of insects, though "rude and accidental" in genesis, is preserved only because of its heritability and adaptive superiority to non-camouflaged insects. Natural selection, Darwin insists, must be part of the equation, indeed much the greatest part, if the lasting variations we do see are to be intelligible. The same, we can read back, must be true with giraffes.

Look again at passages I and II above. The first asserts that understanding why giraffes appear only in Africa and not in other places is about as easy as explaining why some event in human history happened in one place and not another. The second states that some times and places would have been more favorable to the appearance of giraffes than other times and places. Both of these expressions are strongly suggestive of the role of fortuity in human and biological affairs. If pressed Darwin would acknowledge that reasons (causes) exist for all events. But some are so complicated by a vastly large number of contingencies and "accidents" that it would be all but futile even to attempt to give an account of them. Much easier simply to say, much of what comes about is purely accidental; or, if one wishes to avoid the *word* accident (or its synonyms), as Darwin did, one can say instead "we do not know, we cannot know."

If this is correct, then I find no evidence in the passages on giraffes (or in any other passages of Book VII) that Darwin became more Lamarckian in his later years—at least as far as the *Origin* is concerned. He repeated what he had already said in the first edition, that "use and disuse" may assist in some modification, but this is generally to be regarded as a minor role, both in terms of how often it is mentioned in the *Origin* and its relative importance next to other causes. "Direct action of the physical conditions on habits, and thereby on structures," is barely mentioned. In those passages where causes are mentioned, we find essentially the same account as in the first edition, with some refinement and systemization in language. "Chance" (or its equivalents) remains as a fundamental albeit unknown cause, but the terminology is adjusted so as to obscure this point. Above all, Darwin remained thoroughly committed to natural selection as the "most potent" force in evolution through every edition of the *Origin*, and most of the changes he made in the new Chapter VII are best understood in that light.

DARWIN IN 1875

Darwin had little more to say about giraffes after his additions to the sixth edition of *Origin* in 1872. He mentioned them briefly in his *Descent of Man*, published in 1871, but these entries say nothing about his views about variation or adaptation.[49] Nevertheless, he remained focused on the causes of variation in his next major work, *Variation of Animals and Plants under Domestication* (1868), the first part of his "big species book" and the only part to have been published in his lifetime. The title is somewhat misleading, insofar as it suggests he will be dealing only with variation of domesticated productions. In fact he speculates freely, usually by analogy, about variations that occur in nature. The work is interesting and important for showing how Darwin's thinking on the question of variation had evolved in the years directly following the appearance of the *Origin*.[50]

We can deduce very little about his views on variation from the single entry about giraffes in this work, although the fact that he mentioned giraffes at all, considering their slight value for insight about domesticated animals, shows that he was still feeling the sting of Mivart's assault on him for the failure of his theory to be able to account for these marvelous creatures. His concern in the new work was to show that the production of the giraffe need not be the result of sudden massive shifts in structure and organization involving the simultaneous modification of many different parts. Mivart had suggested that the long neck could not exist without the extraordinary supporting structure of the forelegs, and to suppose that such multiple reorganizations of a smaller ungulate all in one blow was beyond any conceivable probability, somewhat like expecting, as one critic expressed it, a perfect poem to emerge from "throwing letters at random on a table" (*Variation,* vol. 2, 207).[51] Darwin was not fazed by such criticism. Like the mammalian eye, the giraffe's neck could be and should be thought of as the result of many small adjustments in one part at a time extending over vast reaches of time. In other words, he reverted to his customary reliance on gradualism and the resource of nearly limitless time, along with the agency of natural selection, to bring about such fantastic results.[52]

But what does the giraffe paragraph suggest about the "proximate causes" of variation (as Darwin had by now come to call them)? Darwin does not give an explicit answer. But, viewed in the larger context of the *Variation* as a whole his comments are revealing. He could have said, and sometimes did say in this work, that variations come about through the "direct effect" of physical conditions on constitution or on habit, and also through the effects of "use and disuse" that sometimes assist variations induced by the former cause. But in relation to giraffes Darwin had only this to say:

> No doubt, if the neck of a ruminant were suddenly to become greatly elongated, the fore limbs and back would have to be simultaneously strengthened and modified; but it cannot be denied that an animal might have its neck, or head, or tongue, or fore-limbs elongated a very little without any corresponding modification in other parts of the body; and animals thus slightly modified would, during a dearth, have a slight advantage [in the struggle for survival] (*Variation*, vol. 2, 206).

This paragraph delivers two messages. The first is that the multiple changes leading to the giraffe were "no doubt" gradual and independent of one another. A more suppressed message, somewhat veiled by Darwin's choice of words, is the suggestion that such slight modifications "just happen," through no assignable cause: neck, head, tongue, or fore-limbs *"might"* just be altered in favorable directions suitable for better prospects in the struggle for life. Darwin's wording points to nature acting on the organism rather than the organism adjusting itself to its environment by "using" its parts to adapt. If this is correct it is far from Lamarck, for whom any such adjustments would be in direct response to environmental needs, not just by happenstance that turned out to be advantageous for the organism in question. Even in his allegedly more "Lamarckian" phase Darwin shows himself to have remained fundamentally committed to random changes plus natural selection rather than use and disuse.

LAWS OF VARIATION IN THE *VARIATION*

Darwin continued to maintain into late life that the "laws of variation" are obscure and often unknown. But he also remained convinced, or at least determined to assert, that variation was a "caused," not a random event. In the *Variation* he attempted again to spell out what could and could not be known about the causes of variation. It is hard to say that Darwin's thinking became clearer in the *Variation*. The *Origin*, by its sixth edition, had managed to reduce all possible causes to five: the *direct* action of conditions on organisms; the *indirect* action of conditions on organisms through alterations of the reproductive system; correlations of growth (not strictly a "cause" but rather a "law," or more precisely several laws, as described earlier); use, disuse, and habit in effecting change; and "so-called spontaneous variation," by which Darwin meant variations that could not be attributed to some other cause or were otherwise unexplainable.[53]

By the time Darwin wrote the *Variation*, however, his thinking had apparently become more complicated. Instead of simply reducing the explanation of variations to discrete "causes," Darwin now presented a more dynamic picture of variation. The main feature of this new account is that one cannot think of causes of variations in abstraction from what is varying. True, conditions of existence, both direct and indirect, have an impact on organization. But now, more clearly than in the *Origin*, variation "depend[s] in most cases in a far higher degree on the nature of the constitution of the being, than on the nature of the changed conditions" (*Variation* vol. 2, 236–40; see also 276–9, 282, and 344–5 for similar formulations).[54] The effect of this shift in emphasis is, if anything, to *reduce* the Lamarckian factor of use-inheritance and to *strengthen* the Darwinian factor of "so-called spontaneous variation." What we need to be thinking about, Darwin now urges, is the "nature of the organism," especially the nature of the reproductive organs of the organisms that produce variations, more than the "nature of the conditions" in explaining variation.

We should not exaggerate, however. It is not likely that Darwin saw any real shift in outlook between the sixth edition of the *Origin* and the *Variation* as concerns causes of variation. Even the doctrine of pangenesis, to be discussed below, while new in detail is not new in basic conception. Even in the earliest editions of the former work we find Darwin confronting an awkward fact for a Lamarckian account. Sometimes organisms of the same species subjected to the very same physical conditions of existence vary a great deal from one another, while conversely, some organisms exposed to sharply differing conditions vary not at all (*Variorum*, 82: 28, all editions; 278–9: 29, all editions; *Variation*, vol. 2, 415; cf. also *CCD* vol. 6, 368, letter to J. D. Hooker April 1857; and vol. 8, 243, letter to C. Lyell, June 1860, in which Darwin calls the inference from changed conditions to variation a "fallacy"). If variation is the first step toward the evolution of new species, these two sets of observations would be hard to explain in Lamarckian terms. For Lamarck, organisms change precisely in response to changing physical conditions. So Darwin was not really breaking new ground, or even altering his causal priorities, when in the *Variation* he emphasized that "the nature of the organism" is more important than the "nature of conditions" in inducing variation. This is where he already stood in the *Origin*.

What, though, about "use and disuse," and whether effects produced by such causes are heritable? Again, setting aside for the moment Darwin's new theory of pangenesis, it is impossible to see greater prominence afforded these factors in the *Variation* than had already been present in the *Origin*. Chapter 24 of the *Variation* is dedicated to examining this cause. The results are predictable for anyone familiar with the *Origin*. Everyone knows, he suggested, that organs and other characteristics may be augmented by use and diminished by disuse. The iron-smith, to use one of Darwin's examples, obviously acquires large arm muscles and other related modifications in the course of his labor. The only interesting question is whether such acquired characteristics are heritable. Darwin addressed this question by asking when specifically variability occurs, before or after conception:

> This is an extremely obscure subject, and we need here only consider, whether inherited variations are due to certain parts being acted on only after they have been formed, or through the reproductive system being affected before their formation; and in the former case at what period of growth or development the effect is produced (*Variation*, 255).

Darwin's answer, developed over the next several chapters, is that inherited variations are formed in both ways, *sometimes* by conditions acting on the living organism and its habits (about this, he says, "there can be no doubt:" *Variation*, 257); and "in many cases" by conditions acting on "the sexual elements, before impregnation has taken place" (a view in support of which Darwin amassed "facts" that "prove" the contention). These are obviously two different answers, the former Lamarckian, the latter perhaps "Darwinian" (if we add a role for natural selection). How can one assign a preponderance of importance between two explanations when comparing "sometimes no doubt" and "proved [to be true] in many cases"? One cannot. Darwin was just where he was in the *Origin*, caught between two worldviews. By 1875 he had abandoned neither, nor decisively moved more toward one or the other.

PANGENESIS

The penultimate chapter of the *Variation* is Darwin's final attempt to bring together under a single theory a number of "grand classes of facts" that cannot be explained by his earlier theory of descent with modification by means of natural selection. He was looking for some way to account for a number of peculiarities in the generation of new forms from parent forms, particularly why new organisms resemble their parents in most ways, or some combination of both parents in the case of sexual reproduction; why offspring display identical developmental patterns as their parents (including the display of similar characteristics at similar times of development and in the same sex); and why in some cases offspring differ from parents (variation). Along the way he also wanted to consider some questions about hybridism, correlation, and reversion. He was here starting to explore the mysteries of reproduction and development. He admitted that his doctrine of pangenesis was "provisional" and "hypothetical," but that a large number of questions needed answers and that pangenesis was at least as plausible as anything else that had been brought forward to that time.

Is pangenesis a "Lamarckian" doctrine?[55] To anticipate, Darwin's new theory *is* in part Lamarckian, but no more so than his earlier theory going back to the first edition of the *Origin*. The basic idea of pangenesis is that each part, or "unit," of every organism consists not only of "cells" that make up the substance of that part, but also of "gemmules," tiny particles within the cells that are "thrown off" from the cells and gathered together in the reproductive apparatus, and are then responsible in some way for the reproduction of the next generation of the part from which they are thrown off. Darwin does not say this, but he seems to suggest that the gemmules contain information—instructions, if you will—that guide organic material in making copies of itself, much like we now regard DNA as doing. Usually the gemmules just make more-or-less exact copies of the tissue from which they were originally "thrown off." Sometimes, however, external or internal events interfere with exact replication, and so offspring do not exactly resemble parents. Many other technical details of this process are offered, but the core idea is that the gemmules reproduce in predictable ways to generate organisms that closely resemble their parents, but that sometimes this predictable course is derailed by extraneous factors that lead to variations.

The question is, what are these extraneous factors? Darwin identified two potential sources of variation. One is that "changed [physical] conditions" have an "injurious effect on reproductive organs," and that such an effect probably causes the gemmules to be "aggregated in an irregular manner" (*Variation*, vol. 2, 388). Such irregularities might thus lead to an increased or decreased size or even abortion of the part from which the gemmules were thrown off. The other cause of variation is the "direct action of changed conditions" on "certain parts of the [already living] body," in turn causing these modified parts to throw off modified gemmules, which are then "transmitted to the offspring" (388). Here is Darwin's summary of the two causes:

> Finally, we see that on the hypothesis of pangenesis variability depends on at least two distinct groups of causes. Firstly, the deficiency, superabundance, and transposition of gemmules, and the redevelopment of those which have long been dormant; the gemmules themselves not having undergone any modification; and such changes will amply account for much fluctuating variability. Secondly, the direct action of changed conditions on the organization, and of the increased use and disuse of parts; and in this case the gemmules from the modified units will be themselves modified, and, when sufficiently multiplied, will supplant the old gemmules and be developed into new structures (390).

This summary is nearly a perfect echo of everything we have heard from Darwin about the causes of variation from the first edition of the *Origin* on, with the added dimension, of course, of gemmules and the doctrine of pangenesis. Sometimes organisms vary because of disturbances to their reproductive organs (how these are induced is not well understood): "The fluctuating variability thus induced is apparently due in part to the sexual system being easily affected [thus causing it often] to fail in its proper function of transmitting truly the characters of the parents to the offspring" (367, reminiscent of Maupertius's theory a century earlier).[56] At other times, living organisms are disturbed by changed conditions of existence, and these changes induce modifications in the gemmules, which in turn lead to changes in the next generation: "But variability is not necessarily connected with the sexual system [but rather sometimes] no doubt results from changed conditions acting directly on the organization [of the living organism] independently of the reproductive system" (367). Why changes of the latter sort should be transmitted to offspring is a mystery: "it is by no means clear why the offspring [of such altered organisms] should be affected by the exposure of the parents to new conditions"; but that they are so affected seems so clear that Darwin concludes, "Nothing in the whole circuit of physiology is more wonderful" (367).

Darwin's two-fold account of the causes of variation expresses the same old dilemma in new dress: do alterations in the germ cells occur before conception, as a result of reproductive organs having been disturbed? Or do they occur after an organism has been conceived (no matter at what stage of development), and thus, through conditions acting on habits causing germ cells to be altered, giving rise to physiological variations determined by these modified germ cells? Darwin embraced both explanations, but he did not change his position on the relative importance of the two causes. Both play a role in evolution, but what weight to assign each role is as it had always been, hard to say. In any case, Lamarckian causes are not augmented.

DEFINITE AND INDEFINITE VARIATION

In light of all of the evidence, then, what is the *best* explanation of Darwinian variation? As usual the question is full of difficulty, but the best theory combines several elements. First, many modifications are acquired during the lifetime of organisms (e.g., the stronger muscles of some artisans: *Variation* vol. 2, 287), whereas others are acquired before birth or even conception (245, 277, 281). Second, these modifications are sometimes heritable (as, for example, longer intestines in some creatures caused by a new diet "must be inherited" (292), although other modifications are not inherited or cannot be known to be inherited (287). In addition, use and disuse are to be seen as *competing* with "selection" to produce lasting variations: "How much of these changes ought to be attributed to mere habit [i.e., use and disuse], and how much to selection of individuals which have varied in the desired manner...can seldom be told" (293). But it was apparent to Darwin that *some* variation must be attributed to use and disuse, and that *sometimes* this variation is heritable: "[All characteristics of plants and animals may] become changed under domestication [and, by inference, under nature], and the changes are *often* inherited" (346, emphasis supplied). His employment of the word "often" in this last sentence more or less sums up Darwin's attitude. "Often" is a modality of most indefinite extension, and can mean anything from "many cases out of a hundred" to "the vast majority of cases." The only things "often" cannot mean are "never," "seldom," and "always."

However, assuming that modalities of relative frequency or significance are important for discerning the degree of Darwin's Lamarckism, two examples may be cited, one from the *first* edition of the *Origin*, the other from the *last* edition of the *Variation*, that, taken together, show where Darwin's heart really was all along.

All of the alterations in his texts that appeared in the intervening 15 years show at most that he was never quite comfortable with how to express himself, not that he had wavering views or that he *ever* was very Lamarckian. In *Origin* 1859 Darwin had this to say:

> I believe that the conditions of life, from their action on the reproductive system, are so far of the highest importance as causing variability (*Variorum*, 117).

That sentiment is quite common in the first edition, and it is not Lamarckian.[57] It is a nice encapsulization of where Darwin perhaps differed most from Lamarck. By 1868 Darwin added a refinement in language by introducing a distinction between "definite" and "indefinite" effects of the conditions of life on organisms. The basic idea is that "definite effects" are those that act in the same way on large numbers of organisms of the same species exposed to the same conditions, making such action to some extent predictable and even understandable. If all members of a flying species grow larger leg bones when placed under domestication (as Darwin believed happened with domesticated ducks over many generations), we may say the effect is "definite," or caused by the new physical conditions. "Indefinite effects" are different. These are variations that differ from organism to organism of the same species, even when the conditions to which they are exposed are the same. Such effects cannot be explained by reference to the Lamarckian factor of "adaptation to conditions," but rather must be explained by reference to "spontaneous variation," or an "unknown cause" (*Variation*, 260).

The question is, which of these two sources of variation did Darwin think was the more common or important? In his "Summary" to the three chapters on the "Laws of Variation" in the *Variation* he leaves no doubt on this score:

> We [earlier] saw that changed conditions occasionally, or even often, act in a definite manner on the organisation, so that all, or nearly all, the individuals thus exposed become modified in the same manner. But a *far more frequent* result of changed conditions, whether acting directly on the organisation or indirectly through the reproductive system, is indefinite and fluctuating variability (*Variation*, 345; cf. 413, emphasis supplied).

There is admittedly some question-begging here: do conditions (in the case of indirect action) act more often *directly* on living organisms or *indirectly* on reproductive elements? Darwin does not say, but in general indirect action of both sorts is less Lamarckian than "direct action." What appears to have happened by Darwin's last published contribution on the question is that he wanted to find an expression that would underscore his preference for the "chance" nature of variations without actually using the expression "chance" or its several synonyms. At some point around 1868 he fixed upon the phrase "indefinite variation." This expression had the advantage of allowing him to acknowledge in one phrase several ideas that always perplexed him: how variations could be so unpredictable, how they could arise independently of physical conditions, and how changes to the reproductive system were really at the base of the great majority of variations. Certainly this would have been the case with the evolution of the giraffe. "Indirect" rather than "direct" variation is where Darwin differed most sharply from Lamarck and where he seems most determined to establish the distinctiveness of his own theory.

NOTES

1. St. George Mivart (1871), *On the Genesis of Species* (London: Macmillan). I follow the pagination in the first American edition (New York: Appleton), 36–41.

2. Authorities who believe Darwin became more "Lamarckian" (explained below) in the years after the *Origin* first appeared include Gavin de Beer, 10–11; P. Vorzimmer (1970, 234–5); E. Mayr (1982, 690–3); E. Mayr (1963, xxiv–xxvii); P. Bowler (1983, 67); P. Bowler (1993, 49); J. Browne (2003, 208, 283–4, 315, 354, 369, 407); *CCD* vol. 11 (1999), 137 n. 6; J. Costa (2009), 494–5. A dissenting view is R. J. Richards (1989), 195 and n. 29.

3. The most accessible edition of the *Origin* is the final one that Darwin edited, the sixth. But some scholars regard the first edition as closer to Darwin's authentic convictions (e.g., M. Ruse [in M. Ruse and R. Richards, eds.], 2009, 1, who claimed that Darwin's revisions in subsequent editions were sometimes made "for less than worthy reasons," i.e., made in response to criticisms that now do not seem valid; cf. E. Mayr [1964], xxiv–xxvii; and J. W. Burrow in C. Darwin (1985 [1859], 11–48).

4. Darwin claims to have seen giraffes in person, as shown in *Descent of Man* vol. 2, 250: "The giraffe uses his short hair-covered horns, which are rather longer in the male than the female, in a curious manner; for with the long neck he swings his head to either side, almost upside down, with such force, that I have seen a hard plank deeply indented by a single blow." Where and when did this observation occur? No doubt at the London Zoological Garden, probably between 1836 and 1838. Owen (1859) reported that four giraffes were brought to the London Zoological Gardens in 1836, and Darwin made visits to the Garden while he was living in London from 1836–39. His visits to London in 1871 may have included visits to the zoo, but these would have been after the reference to giraffes in the *Descent.* A discussion of the details is in Laufer, 1928.

5. This fascinating story is told in Olivier Lagueux (2003, 225–47).

6. Works on Lamarck that should be consulted by the interested student include P. Corsi (1988); R. W. Burkhardt (1977); M. Barthelemy-Madaule (1982); L. J. Jordanova (1984); R. J. Richards (1989); and the Introductions to Lamarck's *Zoological Philosophy* (1984), by R. W. Burkhardt and David L. Hull. Cf. http://www.lamarck.cnrs.fr/bibliographie/biblio_sur_lamarck.php?lang=en for a current bibliography.

7. Darwin added the giraffe's long neck to the third (1861) and subsequent English editions of the *Origin*, in the "Historical Sketch," where the idea of the evolution of the neck is explicitly

associated with Lamarck (*Variorum*, 60: 12). He had already mentioned giraffes in 1857 in a passage intended for (but ultimately deleted from) Chapter 5 of *Natural Selection* (1975, 568–9), in which he suggested that the long neck, tongue, and forelegs could not be explained by Lamarckian factors of use-inheritance, but he did not elaborate.

8. This work, originally published in 1839, is better known today as *Voyage of the Beagle* (1989). Cf. 99–101 in the 1845 edition, reprinted in 1989 by Penguin. It is interesting to observe that A. R. Wallace had also mentioned giraffes, explicitly rejecting Lamarck's account, in his now-famous 1858 paper in which he showed that he had hit upon the theory of natural selection independently of Darwin. The paper was published simultaneously with Darwin's short abstract of his theory in July 1858 in the *Linnaean Society Journal*. Darwin had of course read Wallace's paper prior to writing the *Origin* but he did not mention Wallace's discussion of giraffes in any edition of the *Origin* and it is unlikely Wallace was an influence. Wallace's reference is in de Beer, ed., *Evolution by Natural Selection* (1958, 277).

9. A "Google Books" search of "Darwin giraffe 1860–1882" yields hundreds of hits, but rarely do these get at the evolutionary significance of these animals for Darwin. The exceptions are typically reviews of Darwin's *Origin* that appeared after Mivart's critiques in 1869–1870 and were prompted as much by Mivart as by Darwin. The most famous example is perhaps the battle of words that broke out in 1871 between Chauncey Wright, who argued that Darwin's theory could explain the giraffe's neck, and Mivart himself, who continued to maintain it could not (most of the reviews of the time that I have seen supported Mivart). Cf. C. Wright (1871); and St. G. Mivart (1876). More recent entries are usually popular reconstructions and pay little or no attention to textual or theoretical concerns. Some exceptions are: F. Hitching's anti-Darwinian argument (1982); S. J. Gould's rebuttal (1996, 22); C. Pincher (anticipating some of Gould's argument) in an article that appeared in *Nature* (1949), suggesting that the legs, not the neck, were "selected," and that the neck was a necessary compensation. Giraffe biology is well covered in A. I. Dagg and J. B. Foster (1976).

10. S. J. Gould (2002), 188 reminds us that giraffes were perhaps not as important to Lamarck as is often believed. He mentioned them only once in his *Philosophie Zoologique* (1809, 122, translated into English in 1984 by H. Elliot as *Zoological Philosophy* [hereafter *ZP*]), and is said later to have regretted bringing them in at all (cf. A.S. Packard [1901], 411 n. **, citing works by the French Lamarckian A. Giard). Modern commentators generally agree with Gould. Burkhardt, for example, thinks giraffes for Lamarck were probably "an afterthought" (1977, 2) and that the giraffe has sadly remained the only example of "Lamarckism" to have survived in the common understanding, and then only as a caricature of his thinking (201). Bowler (1983, 62–3) and (1993, 3–4) nevertheless does employ the giraffe example to illustrate Lamarckian "use-inheritance."

11. Giraffes illuminate only part of Lamarck's theory. The editors of Darwin's correspondence (*CCD* vol. 11, 231, n. 12 [March 1863]) describe his theory as consisting of "two linked factors," the first "a natural tendency toward organic complexity" (called by Lamarck the "power of life" and by Darwin "progressivism"); the second "the direct influence of the environment on the organism" (often called "adaptationism"). Giraffes bear only on the second factor, whereas Lamarck regarded the first factor as more important. Cf. Corsi (1988a, 194, 197, 243–5); Burckhardt (1977, 144–51); Bowler (1983, 7, 42–3, 58–63); Jordanova (1984, 54–5); R. J. Richards (1989, 49–50). Darwin was always clear that he rejected Lamarck's "progressivism" (e.g., *CCD* vol. 11, 222–3 [March 12–13, 1863], letter to Charles Lyell; *Marginalia* [1989, 479]: "Here is the difference between Lamarck & Me").

12. At first Darwin downplayed the significance of the difference between his own theory and Lamarck's. In *Natural Selection* (1857, published in 1975) he wrote: "It is immaterial to us, whether slight changes of structure supervene first leading to changed habits, or whether habits and instincts change slightly first, & the animal is benefitted in subsequent generations by slight selected modifications of structure in relation to the already slightly changed habits" (340). In fact the distinction between these two views goes to the heart of the difference between Lamarck and Darwin's theories. Lamarck always insisted that a change in "habit" preceded a change in structure, whereas Darwin came to insist in the *Origin* and thereafter that changed structures as a rule precede changed habits.

13. Darwin confessed to Bunbury in 1845 that he believed in "transmutation," but not the Lamarckian variety—an early public confession to someone outside of his close circle of friends (*CCD* vol. 3, 237, n. 5 [July 1845]). Cf. R. J. Richards (1989, 86–9; 92–3).

14. Darwin's son Francis commented in his "Introduction" to his edited volume containing the 1842 "Sketch" and 1844 "Essay" (reprinted in *Evolution by Natural Selection*, hereafter *EBNS*), edited by Gavin de Beer (1958), that Darwin did not need Malthus to see that he had a theory of his own and expressed surprise that Darwin would have given Malthus much credit for helping him to crystallize his ideas (*EBNS* 27).

15. Ghiselin (1969, 48–9, 59–77) argues (see 256–7, nn. 14–17) that Malthus was important for getting him to think in "populational" terms, important both methodologically and substantively for the theory. R. J. Richards (1989, 98) appears to agree with Ghiselin. But cf. Hodge and Kohn (Kohn, 1985, 194–5); and D. Kohn (1980).

16. Lyell eventually came to see that Darwin was not a "progressionist" in Lamarck's sense, as he indicated in his 1863 *Geological Evidences of the Antiquity of Man*, 412. By this time he had come to count Darwin's departure from Lamarck on that point as a positive development, but he had not always seen it. Cf. J. David Archibald, *Ladders, Circles, Trees: Changing Visualizations of Our Place in Nature from the Greeks to DNA* (Columbia University Press, forthcoming); S. J. Gould (1981).

17. Cf., e.g., Moore (1979, 142–3); Burkhardt (1977); Corsi (1988); S. J. Gould (2002).

18. Lamarck, in this part of his theory, was also thought (by Darwin and others) to allow an apparent role for "conscious willing" on the part of organisms to change their habits, and thereby structures, to changing "needs" ("*besoins*"). When Darwin departed from Lamarck's account of adaptation, he also departed from this particular part of the account. Finding a role for "conscious willing" in Lamarck is now generally regarded as a mistake (cf. R. J. Richards, 1989, 92–3; Shanahan, 1991).

19. Cf. Darwin's strictures on "higher/lower" (e.g., *Charles Darwin's Notebooks*, hereafter *CDN*). *CDN* E-95–6: "why not dis-development...law of development in partial classes far from true." *CDN* E-68–9: *CDN* N-47. Cf. Sloan, 1986, for Darwin's early recognition that less organized forms are likely to have longer "species-lives" than more complex forms; and Ospovat, 1981, for Darwin's shift from a "perfectionist" understanding of adaptation to a "better than others" view.

20. *CCD* vol. 3, 253 (letter of Sept. 1845 to J. D. Hooker); cf. *CDN*, 276 (C-123), 287 (C-157–8); 275 (C-119, wherein Darwin called Lamarck "the Hutton of Geology, bold in many such profound judgment—the highest endowment of lofty genius"). For discussions of Darwin's various encounters with Lamarck's writings, cf. S. J. Gould (2002, 196–7); F. N. Egerton III (1976, 452–6); R. J. Richards (1989, 86–90); and J. Browne (1995, 75 ff).

21. J. B. Lamarck, *ZP*, 110–18, and cf. *Origin* [*Variorum*], 83, 282–3; *Variation* vol. 2, 207, 248, 306–9; cf. Burkhardt (1977, 182). Lamarck, in turn, apparently took the example of domesticated ducks (Darwin's favorite example of the effects of use and disuse) from J. J. Virey's ornithological studies (cf. Corsi [1988], 98).

22. From an early time in his reflections on origins Darwin was sure that Lamarck was wrong, at least in some important ways, and that his ideas were very different: "<<my theory very distinct from Lamarcks>>" (*CDN*, 224: 214 and n. 1 [1837–8]); *Marginalia* (1990), 477–80. Other references to Lamarck may be found in Darwin's *Autobiography* (1958), 49; and in correspondence to his close friends J. D. Hooker, where he said "heaven forfend me from Lamarck nonsense of a 'tendency to progression'" (*CCD* vol. 3, 2), and Charles Lyell, to whom he wrote "[Lamarck's *ZP* is] a wretched book from which (I well remember my surprise) I gained nothing" (in F. Darwin, 1887, vol. 2, 198–9). Cf. *CCD* vol. 7, 348.

23. As early as 1837, Darwin expressed scorn for Lamarck's doctrine that animals can somehow will changes to their structures: "Lamacks [*sic*] 'willing' doctrine absurd" (*CDN* B-216; cf. *CCD* vol. 3, 79 [November 10–11, 1844, and nn. 12–13] to Hooker: "Lamarck's [book is] veritable 'rubbish' [on insects "willing"]"; vol. 3 253 [September 1845] to Hooker: "[Lamarck (is) absurd though clever]; and often). Cf. also Darwin's marginalia on Lamarck 1830: "it is absurd the way [Lamarck] assumes the want of habit cause annihilation of organ & vice versa" (*Marginalia* [1990], 480, written in 1839). An early twentieth-century defender of Lamarck, A. S. Packard (1901, 351–3), argued that Lamarck had been completely misunderstood by Darwin and others, that the "willing" was never intended to be a conscious striving but rather only a "physiological" reflex or instinct. Corsi (1988a, 196), and especially Burkhardt (1977, 175) give credence to Packard's understanding. But misunderstanding or not, the idea that creatures somehow strive or endeavor to change their habits has persisted into modern times as the correct way to understand Lamarck (Richards [1989], 51–2 and n. 99; 92–4).

24. Use-inheritance is important for identifying Lamarckism (Burkhardt [1977, 1–3, 144]; Gillispie [in Glass et al., 1959, 265–91]; Mayr [1982, 689–93]); Bowler (1983, 62–6); Richards (1989, 92–4 and often). Darwin could easily believe, as anyone can, that organisms respond to environmental influences (consider getting a suntan) and that organs may be augmented or diminished by use or disuse (the enlarged arms of the iron-smith). The key Lamarckian question is whether such changes are transmitted to offspring. If they are not they are unimportant for evolution, as Darwin emphasized (*Variorum*, 85: 49).

25. Darwin's formal position in the first edition of the *Origin* was that use-inheritance may account for *some* variation (how much is not clearly spelled out because other factors are also in play), but that the really important point is that *no* variations will be retained over time without the "predominant" role of selection (natural or human) to preserve variations that do arise from whatever cause (*Variorum* 119: 322; 290: 96; 320: 305). This point is emphasized in his letters (cf. *CCD* vol. 8, 176–7, 179, n. 4 [April 28, 1860] to Andrew Murray; vol. 13, 91 [March 25, 1865] to Lyell).

26. Darwin had actually performed experiments on "use" as affecting ducks' physiology as early as 1854, as he reported to Hooker (vol. 5, 250 [December 11, 1854]; cf. vol. 5 352 [June 11, 1855] to Fox). C. Darwin, *Variorum*, 83: 34 (retained in all editions); *Variation* vol. 2, 288, 367; cf. 307.

27. Buffon had believed that cropping the ears and tails of puppies would, in time, "transfer those defects...to their descendents" (in R. J. Richards [1989], 37). Late in the nineteenth century another Darwinian, August Weismann, used mice to test this Buffonian belief by cropping

tails through many generations. He failed to produce tailless progeny, helping to confirm his conviction that use-inheritance is no part of evolution. The post-Darwin fate of "Darwinism" and the disappearance of Lamarckian elements is told in Bowler (1983) and Richards (1989).

28. The examples are taken from *Variation*, vol. 2, 287–90, 390–2. Darwin probably borrowed the circumcision example from J. Wyman, who had written to him about this in 1860 (*CCD* vol. 8, 360 [September 1860]).

29. One may ask whether Darwin's thoughts about "use-selection" had undergone any noticeable change *prior to* the first edition of the *Origin* in 1859. Again the subject has generated controversy. For example, T. H. Huxley, on reading in 1887 Darwin's unpublished "Essay" of 1844, thought that Darwin had assigned "much more weight" to use-inheritance in 1844 than in 1859 or thereafter; whereas Gavin de Beer, who edited the "Essay" for publication (1958, 37) thought the opposite. In 1857 (*Natural Selection* 365) Darwin was quite emphatic that "the effects of [use/disuse and] habit were of quite subordinate importance [to natural selection.]" Later in the same work the Lamarckian doctrine of use/disuse as the sole cause of change was deemed by Darwin to be "utterly false" (368).

30. For the impression that Darwin thought he had become more Lamarckian after 1859, cf. *CCD,* vol. 10, 556 (October 1862), letter to Hooker. Then in 1875, to his cousin Francis Galton he wrote, "Every year I come to attribute more and more to [the Lamarckian] agency [of use and disuse] in accounting for modifications"; and by1878, looking back to the *Origin*, he wrote, "I probably underrated [the Lamarckian] power [of external conditions] in the earlier editions of the *Origin*" (both quoted in Mayr, 1982, 691). Cf. also Darwin's letters to Mivart in early 1871 (*CCD* 19, 30–7), in which Darwin defends himself against Mivart's accusations that Darwin assigned "too much weight" to natural selection. For the contrary impression, cf. F. Darwin, ed., (1959), vol. 2, 514–5 (letter to J. H. Gilbert, February 16, 1876); 516–17 (letter to K. Semper, July 19, 1881).

31. See previous note. "Highly important" looks like an exaggeration here. In 1860, when the first reviews and translations of the *Origin* were coming out, Darwin often objected to suggestions that he had attributed too much to use-inheritance. For example, to Bronn (the German translator), Darwin wrote that the rendering of "natural selection" as "Wahl der Lebens-weise" sounded too "Lamarckian," giving the impression (that Darwin here rejected) that "habits of life [are] all important." The emphasis, he wrote, should be on "selection." (*CCD* vol. 8, 83, February 14, 1860). Similar in message are comments to other correspondents (e.g., vol. 8, 176–7 and 179, n. 4 [April 28, 1860], to Andrew Murray; vol. 8, 193 [May 7, 1860] to T. H. Huxley; vol. 8, 243 [June 6, 1860] to C. Lyell). On the other hand, some reviewers thought Darwin had attributed too *little* to use/disuse (cf., e.g., *CCD* vol. 8, 100 [February 23, 1860] from J. Lamont; vol. 12, 391–3 and n. 8 [November 3, 1864] to Hooker; H. Spencer, *Principles of Biology* [1896], 449).

32. Typical in this regard is a change Darwin made in the fifth edition about how to characterize the operation of natural selection vis-à-vis other factors. In editions 1–4 he had written, "Natural Selection has been the main but not the exclusive means of modification." In the fifth edition this became, "Natural Selection has been the *most important*, but not the exclusive means of modification" (*Variorum*, 75: 50 and 50e, emphasis added). A small change no doubt, but one that, if anything, strengthens the role of natural selection.

33. Lamarck was not the first naturalist to have shown an interest in giraffes. According to Cuvier, both Linnaeus and Buffon had noted the peculiar features of this animal and had tried to find systematic ways to incorporate it into their classificatory schemes (G. Cuvier, *Animal*

Kingdom, 1834–37, vol. 2, 170). But Cuvier did not discuss the sources or causes of its remarkable adaptations; presumably he thought giraffes were simply created that way. Lamarck was apparently the first one to give a transmutationist account.

34. Burkhardt (1977, 168–9) gives an excellent account of Lamarck's often misunderstood *sentiment interieur*, in which he properly emphasizes the point that conscious "willing" on the part of organisms is often no part of the interior sentiment.

35. To effect change through effort an organism must be alive, but not usually adult. Lamarck thought that younger creatures were more likely to be able to effect change in organization through effort than older ones because of the greater plasticity of parts in younger organisms (Corsi, 1988a, 131).

36. Mivart's reviews were gathered together in a separate volume *On the Genesis of Species*.

37. Lyell (1863, 410) gave a brief account of Lamarck's discussion of giraffes in *ZP*, but only to say what he believed Lamarck had argued, not to criticize or comment on Lamarck's account.

38. Elsewhere in the *Origin* Darwin did try to enumerate and explain the "laws of variation," even devoting an entire chapter (Chapter V) to this very subject, and among the "laws" he mentioned, "use and disuse" was one, but always a secondary or subsidiary one (cf. n. 24 above; and chapter 3 of the present work).

39. E. Geoffroy St. Hilaire also believed that changes of structure precede changes in habits in the order of evolutionary change ("Geoffroyism"). But for Geoffroy the changes in structure are induced directly on living organisms by changes in environment, whereas for Darwin, the most important cause of changes in structure is the *indirect* action of the environment on reproductive elements. Cf. E. Mayr (1983, 687).

40. The "Selection" of this passage, which comes at the end of Chapter I, is human rather than natural selection, but the sentiment, though more clearly expressed here than anywhere else in the *Origin*, remains the same in regard to natural selection throughout. The one slight concession in this passage made to his critics in the sixth edition was the replacement of the definitive "is by far the predominant Power" to the more tepid "*seems* to have been the predominant Power." This is a good illustration of the sort of change Darwin typically made to the sixth edition, apparently feeling the heat of Mivart's attack. He did not remove Natural Selection as the most potent force in evolutionary change, but he threw in some qualifiers that might seem to enlarge "use" at the expense of "natural selection" but really do no such thing.

41. Of the three passages quoted here Gould (1996, 23) notices only the latter two, and so misses the point I am making here, based on the explicit wording of the first passage, that Darwin saw "use" only as "aiding" variations that originated through other means. In our example, the *primary* cause of the longer necks is not given by Darwin, but he does allow that once the longer necks have appeared, repeated use may "aid" in making the necks longer still. The same idea appears several times in the *Variation*, e.g., vol. 2, 305, 308, 415. As Burkhardt (1977, 177–8) has pointed out, attributing enlargement or diminishment of *existing* organs to use and disuse respectively was easy enough. Explaining the *origin* of new structures was the hard nut, and Lamarck's explanation was "feeble." Darwin in the giraffe passages accepted the first but never the second. Use "aiding" a variation that has already occurred through other causes would be a *secondary* cause (cf. Vorzimmer [1970, 14] for the "primary/secondary" distinction, wherein "secondary" causes are incapable by themselves of bringing about new features).

42. "Use" aiding prior variation was already Darwin's position in 1857 when he wrote *Natural Selection* (only published in 1975): "I doubt whether anyone would have thought of training a

dog to point, had it not first shown some *innate propensity* in this line.... In this case the selection of a *self-formed propensity* would have had as much to do with the formation of the breed as habit" (484, emphasis supplied).

43. Darwin rejected the role of "will" in effecting change (assuming, incorrectly, that Lamarck had accepted it—cf. n. 17 above). Yet "use" almost implies willing effort, as Darwin seems to say, in spite of himself, when speaking of fish that lay on their sides and have two eyes both pointing upwards: "We thus see that the first stages of the transit of the eye from one side of the head to the other... may be attributed to the habit [of the individual] of endeavoring to look upwards with both eyes, while resting on one side at the bottom" (*Variorum*, 250–252, especially VII. 382.65.0.50.158). The example, as often in Darwin, was taken from Lamarck [*ZP*, 120], but the "endeavoring" is Darwin's addition. The flatfish example, interestingly, had been noted by other naturalists prior to the sixth edition of the *Origin*, and Darwin was clearly influenced by these other references. A. W. Malm (1867, cited in *CCD* vol. 14, 240, n. 3) had discussed this issue, and Darwin treated his views at some length (the note appears in an annotation to a letter to Darwin from Charles Kingsley, who asked Darwin to explain the two dorsal eyes in Flatfish; the editors of *CCD* refer the reader to *Origin* sixth edition 186–8, concluding that Darwin believed the phenomenon to be "the result of inherited effects of use, possibly aided by natural selection"); and so had St. George Mivart (1871), 75, who found this example to speak strongly against natural selection because the slow migration of one eye could hardly be beneficial during transit.

44. *Pace* Browne (2002, 311), the idea that some variations have no adaptive purpose was *not* new with Darwin in the late 1860s, after he had read Mivart and Jenkin. One may trace it to some of his earliest musings on variation in the *Notebooks* (e.g., *CDN*: 633–6), work leading up to the *Origin* (*Natural Selection* 214), and it stayed with him through the intervening years (e.g., *Variorum* 164: 14).

45. This account squares well with that offered in Vorzimmer (1970, 236) that variations can be augmented but not caused by use and disuse: "And here Darwin saw that the continued or *sustained* action of conditions could both amplify and extend the very modifications which it [variation] had initiated" (Vorzimmer, 1970, 236).

46. Chapter 6 of the present work traces Darwin's shifting stance on "chance" as a cause of variation and the substitution of "so-called spontaneous variation" as a stand-in for chance. The latter expression appears more than a dozen times, starting with the fourth edition (examples can be found at *Variorum* 232: VII.382.61.0.0.5; 235: VII. 382.61.0.12.8; 253: 382.65.0.50.166; 364: 190.f; 369: 219+20.f; and 382: 28.e), whereas the word "chance," employed more than 10 times in the first edition as a cause of variation, almost disappears completely by the sixth edition.

47. Darwin's deployment of insects in such direct juxtaposition to giraffes was actually a clever move. It would be easier for Mivart and his supporters to believe that insect camouflage could be explained in Darwinian terms than for them to believe giraffes could be so explained, mainly because of the difference in size (and presumably complexity) of the two sorts of creatures. One of Mivart's objections to Darwin's theory was that, in the case of giraffes, so many simultaneous and coordinated variations would have to occur to effect structural change as to render the event improbable in the highest degree. This is not so clear in the case of insects. Yet, Darwin is saying, the two cases are essentially the same.

48. Mivart, *Genesis of Species* (1871, 32–3, 82–3). Cf. Vorzimmer (1970, 231–3).

49. *Descent* (1981, [1871], vol. 2, 250, 271).

50. We should note that some overlap exists between the last two editions of the *Origin* (1868 and 1872, respectively), and the first edition of the *Variation* (writing began in 1862, first published in two volumes in 1868). A careful comparison shows Darwin altered the final two editions of the *Origin* to conform to his new wording about variation in the *Variation*.

51. Darwin was intrigued by how apparently "fortuitous" events in nature were ultimately resolvable into deterministic "laws." Just as one can predict, with sufficient knowledge of the laws of physics and other parameters, how letters thrown randomly on a table will fall out, so too one can predict how a handful of "feathers [thrown up] on a gusty day... would fall," or how a chemist who "throw[s] a dozen salts into a solution may hope to predict the results" (*Natural Selection*, 184, 198). Even as early as 1838 Darwin had decided that both the "throwing up of a farthing [and how it lands]" and "our throwing it up" (viz. the motives of the thrower) are both "determined," i.e., not due to "chance" (*CDN* M 27).

52. Cf. *Variation* vol. 2, 327–9 for Darwin's response to Herbert Spencer's criticism (1896), 450–6, that the enormous set of antlers on the Irish Elk could not be explained by natural selection because of the multiple other adjustments in organization that would be necessary to support such a structure. Again, Darwin defended natural selection of incremental changes. Cf. Mark Ridley, "Coadaptation and the Inadequacy of Natural Selection" (1982, *BJHS* 45–68); R. J. Richards (1989, 291–2 and n. 156).

53. With some effort one can reduce these causes/laws to three: *direct* action of conditions on organisms includes the Lamarckian factors of "use, disuse, and habit" and the Geoffroyian factor of structural changes directly induced by the environment (e.g., cold climate inducing longer hair) without the mediation of "habit;" *indirect* action is the same as conditions acting on reproductive organs that in turn induce variations "spontaneously;" "correlations of growth," itself involving multiple laws, remains the same from the *Origin* to the *Variation*.

54. Darwin added this same statement, almost *verbatim*, to the sixth edition of the *Origin*, but as he acknowledged the idea had first made its appearance in the *Variation*, first edition, that had come out four years before the sixth edition of the *Origin* (*Variorum*, 78: 9.2).

55. That it is seems to be an accepted view in recent scholarship. Cf. J. Browne (2003, 354); Mayr (1982, 693–4); Bowler (1983, 65, 78); Bowler (1993, 31–2).

56. Cf. B. Glass in B. Glass et al., eds. (1959, 68, 81).

57. Other instances appear at *Origin* (*Variorum*), 79–82: 12–13, 20, 28; 87: 63; 166: 22; 276: 8; 279: 30–32e.

9

Chance and Free Will

THE M AND N *NOTEBOOKS*

In his so-called "Metaphysical Notebooks," *Notebooks* M and N, Darwin started in earnest to explore how his theory would necessarily force a rethinking of some of the most settled convictions of scientists and non-scientists alike about how and why humans are different and special. Of central concern were issues of morality. To be fully "moral," it was widely agreed, humans would need to possess free will (cf. Richards, 1989, 240–1). They would need to be equipped with the sort of intelligence that would enable them to choose right from wrong. They would need to be the sort of creatures who were in command of their own moral destiny. They would need to be able to show themselves to be meritorious in the eyes of an omniscient and good God. If humans, like all creation, are merely "by chance," what becomes of these various needs?

Darwin might have argued at this point that even if humans have evolved by "chance" variation they are still capable of exercising free choice and making morally correct decisions. To say that something has come about through "chance," in other words, does not preclude one necessarily from believing that thing to be capable of freely chosen morally correct or incorrect conduct. In Darwin's world, where the existence of an omniscient God was widely assumed, such an argument would be more difficult than saying humans arose through divine agency precisely for the purpose of being morally worthy creatures, but it would not be impossible. Indeed, some secular humanists and other moral philosophers who reject the existence of God often bring forward just this sort of argument to defend the possibility of human goodness in a godless world. This is the path taken by some prominent Darwinians today, including Richard Dawkins (1989, 2006) and Daniel Dennett (1995, 2006). They prefer a "soft landing" to the hard crush of "survival of the fittest," at least for humans. But that is not the path Darwin took.

Instead, Darwin used his understanding of chance to wage an assault on the very possibility of free will, which he correctly understood to be a fundamental premise of most of the moral schemes of his day.[1] Free will, he argued, is a lot like chance; in fact it means the very same thing. "I verily believe free will & chance are synonymous," he wrote in *Notebook* M (M-31). What did Darwin have in mind in saying this, and what were the consequences?

A preliminary: If "chance" and "free will" are synonymous, and if both are illusory, what becomes of the premise of this book? I have been arguing all along that "chance" was *not* illusory to Darwin, and therefore, by implication, neither can be "free will" on Darwin's terms. Either "chance" and "free will" are imaginary constructions of gullible human minds (and so are illusory), or both are "real facts" that may be studied empirically, to be verified or disconfirmed by careful observation. We must confront the possibility that Darwin was inconsistent.

The apparent discrepancy may be overcome by recalling arguments already made. "Chance" as the framing concept of our discussion did not mean to Darwin only "cause unknown." It also meant "no assignable reason." Free will should be regarded in the same light. Free will in human behavior does not mean "uncaused." To Darwin it meant "the reasons for actions that are usually attributed to freedom of will" may more accurately be understood as "determined by reasons (motives) that are not always immediately assignable, but that ultimately may be resolved into deterministic laws." So, to abbreviate, "chance" and "free will" are synonymous, not because no reasons or laws can be assigned for their operation, but rather because such reasons and laws are often beyond human comprehension; and also, because such reasons and laws cannot plausibly be assumed to be directed by divine intelligence. "Free will" and "chance" come under the same deterministic principles—hence, they are "synonymous."

FREE WILL AND CHANCE

To elaborate, we may "fancy" that we have free will, just as we may fancy that chance governs events in the universe. But in both cases something else is at work. The illusion is created by the peculiar operation of the human mind when it confronts phenomena of a particular sort. In the case of free will, the mind watches itself deciding, making choices, preferring this act to that act, and so forth. Because the mind does not detect any impediments to its ability to think and to choose, it infers "free will" as the psychic mechanism that enables it to do these things.

Darwin argued instead that the thoughts that give rise to these apparent choices are themselves not "free," but rather are "determined," by heredity and upbringing. The full theory, if spelled out in its particulars, goes something like this. Humans, like all other animal creatures, are born with mental equipment inherited from their parents, and this equipment already predisposes humans to have some preferences rather than others from the start. During development new dispositions and appetites are acquired through experience and instruction. (Darwin apparently did not extend his reasoning to plants, on the grounds that plants have no cognitions whatsoever. For plants, therefore, the question of free will does not arise.) Choices, the argument continues, are nothing more than the result of assessments about what one's current appetites are for and how best to satisfy them. But since one does not "choose" one's appetites (e.g., no one "chooses" to be hungry or tired or sexually aroused), one does not freely "choose" actions that presumably lead to the satisfaction of those appetites. Instead one merely calculates what actions are most likely to produce the satisfaction of already existing appetites. The "choice" in each case is determined by the appetites that happen to exist at any given moment.[2] One might paraphrase here by saying that the appetites that govern choice are "by chance," in the sense that they are themselves not chosen. They come about through processes that are entirely beyond the rational control of cognitive decision-makers—heredity and upbringing. In Darwin's words:

> [That] affections [are the] effect of organization...can hardly be doubted.... The common remark that fat men are good-natured, & vice versa.... Thinking over these things, one doubts existence of free will; every action determined by hereditary constitution, example of others or teaching of others.—(NB man much more affected by other fellow-animals, than any other animal & probably the only one affected by various knowledge which is not hereditary & instinctive) and the others learnt. What they teach [comes about] by the same means & therefore properly no free will.—We may easily fancy there is, as we fancy there is such thing as chance.—Chance governs the descent of a farthing, free will determines our throwing it up.—Equall[y] true the two statements (*CDN* M-27).

At first glance one might wish to conclude from this passage that Darwin believed humans are special—(e.g., "the only one[s] affected by various knowledge which is not hereditary"). In the larger context, however, we see that what Darwin is asserting is that choice, for example the choice of being "good-natured," is tied to hereditary constitution in all creatures, but also that "knowledge" (i.e., what one is taught after heredity has had its say) can alter hereditary effects, and does so especially and perhaps uniquely in "man." But such acquired knowledge does not activate or enable a "free will" not present in other animals. It merely changes the genetic program. The only plausible meaning of "free will determines our throwing [the farthing] up" is that the choice of the thrower to toss the farthing has been "determined" by what he has "learnt," rather than by his hereditary constitution alone. In just the same way, one must infer, the question of whether the farthing lands heads or tails must depend on laws of physics, as much as we might "fancy" the outcome is due to "chance." Both cases—the tossing and the result—are determined by factors that allow no room either for choice or chance.

The idea may be illustrated by considering the tossing of a coin, or of dice. Forget for the time being the motives of the one tossing. We shall come back to this. Just assume a person tossing a coin or dice. Darwin's point is that the outcome of the toss or the roll, no matter how much we wish to say the result is purely chance, is not really so at all. The result is governed by laws of nature. One may extrapolate: if a person tossing coins or dice threw in exactly the same way, with exactly the same coin or dice, from exactly the same starting position, in exactly the same environment, and with precisely the same governing conditions, the results of the toss would be exactly the same. What leads us to say the results are "by chance" is simply the result of the governing conditions being, as a general rule, not the same, making the roll or the coin-flip unpredictable. "Chance," on this view, is just shorthand for saying that, since the governing conditions can never be assuredly the same in each roll or toss, the outcome can never be assuredly the same. The result, in consequence, inevitably appears fortuitous. But it is not. It is dependent on the various conditions within which the exercise takes place and the various laws of physics that determine how coins or dice behave when they are tossed.

Matters may seem to be complicated when one introduces a human agent who makes choices about when, where, and how to flip a coin or to roll the dice. But for Darwin nothing changes. Humans who decide to engage in such activities are just as much bound by the laws of their own nature as are the coins and the dice bound by the laws of physics. In other words, "free will," the supposed interloper in an otherwise deterministic world, is just another pre-programmed role-player. Decisions made by human actors are just as much a result of the "organization" of the human psyche as the "decisions" made by oysters to clamp down when confronted with danger. Darwin saw no real distinction.

> I verily believe free will & chance are synonymous.—Shake ten thousand grains of sand together & one will be uppermost:—so in thoughts, one will rise according to law (*CDN* M-31).

And then again much later in *Notebook* M:

> The free will (if so called) makes change in bodily organization of oyster, so may free will make change in man.—the real argument fixes on hereditary disposition & instincts...My wish to improve my temper, what does it arise from but organization. That organization may have been affected by circumstances & education, & by choice which at that time organization gave me to will.—Verily the faults of the fathers, corporeal & bodily are visited upon the children (*CDN* M-73).[3]

Choice does not determine a person's order of preferences including, by extension, a person's "choice" to "improve [his] temper," any more than "chance" determines which grain of sand comes out on top or an oyster's "ability" to change its organization. Rather, the order of preferences determines choice. That order is as bound—now explicitly by "law"—as how sand when shaken will sort itself out among grains on top and those below and how oysters undergo modification of structure over the course of ages. In the latter cases we assume Darwin has in mind laws of physics and biology. What law, then, might govern the ordering of human preferences and so the choices they make? And if human choices are "determined" by laws rather than by human will, what becomes of a morally ordered universe presided over by a wise God?

The answer to the second question appears to be that the belief that the universe is morally ordered by divine decree must be abandoned. The idea that if humans have free will then so do puppies, oysters, even "polyps" leads to an absurdity. Rather than allow the latter, Darwin dismissed free will and all its associated baggage even from the human realm. He again came back to his earlier conclusion that the existence of free will is illusory. It is no more justifiable to believe that man has it than any other creature. What governs animal behavior through and through—including humans—are motives determined by heredity and experience. Darwin's view, he confessed, was a new kind of predestinarianism:

> The real argument [about how humans make choices] fixes on hereditary dispositions and instincts.... The above views would make man a predestinarian of a new kind, because he would tend to be an atheist.... It may be doubted whether a man can intentionally wag his finger from real caprice. It is chance which way it will be, but yet it is settled by reason (*CDN* M-74).

One sees that Darwin has come to the nub for moral philosophy. To eliminate free will is to invite atheism. The "predestinarianism" in the passage is not the old kind, according to which an omniscient Creator has a plan for the universe that He foresees will be fulfilled, despite human depravity. It is rather a "new kind" of story about human life, one that is absent a Creator altogether, but nevertheless one that unfolds according to its own irresistible logic.

And what is this logic? To put the question differently and in explicitly Darwinian terms, what are the "laws," alluded to above, that govern human choice? "Free will" implies the absence of natural laws for regulating human conduct, for a law is a fixed rule that cannot be overcome by "caprice," whereas if humans were truly free, deterministic laws could not settle their behavior, by definition. Lyell and others made this point explicitly to Darwin (*CCD* vol. 8, p. 260 [July 19, 1860]). Either law or free will would have to go. By its nature free will can and does intervene to overcome deterministic laws; if it could not do so, it would not be "free." Darwin, as already seen, took his stand with deterministic laws against free will. Free will has no meaningful role to play in human conduct, except perhaps by befuddling human understanding. But by retaining "laws" as the correct answer, Darwin was confronted with additional questions: how to describe these laws, and how to explain *their* origin.

As to the latter, Darwin did not give an unequivocal answer, and that fact made it possible for theologically inclined readers (including friends Lyell and Gray) to believe Darwin was still within Christian respectability. If God made the laws to operate in just the way they did, evolution by natural selection could be adopted as a respectable scientific theory. But what if no God presided even over the fashioning of the laws? What if the laws were really just by "chance"? That answer would present an obstacle that even supporters would trip over. I have discussed this issue in earlier chapters. I shall summarize here by saying only that Darwin did not and could not commit, except in the privacy of his own brain. His public statements were always to the effect that the question is too profound.

HOBBESIAN MORAL PHILOSOPHY

To answer the question about how to describe the laws that govern human choice, Darwin reverted to an account that goes back to the British philosopher Thomas Hobbes, and which Darwin may have learned of from David Hume. But before giving a summary of Darwin's analysis, a short digression into Hobbesian moral philosophy will be helpful.

Hobbes also had confronted the question of human free will, admittedly in a very different context from that of Darwin, but in a context, nevertheless, where the role of law in human behavior had to be reconciled with a role for free will. According to Hobbes's account, to abbreviate, human beings, living in a "state of nature" (another familiar Darwinian expression, but one that for Hobbes meant essentially a "condition" in which humans live without governments or man-made laws), are confronted with the need to decide what rules (i.e., laws) to obey if they are to survive the "war of all against all." Do such laws exist even in the absence of governments? (This "war of all against all," incidentally, sounds very much like the Malthusian "struggle for survival" in a "natural" context of competition for scarce resources among equally acquisitive beings. Malthus, too, was influenced by Hobbes's thought.) Hobbes reasoned that humans, being rational creatures, would come to see that survival would depend upon each individual observing certain "natural laws," or, as he called them, "Laws of Nature." Without such observance the war of nature would result in a life for each that would be "solitary, poor, nasty, brutish, and short." In other words, life for humans in a state of nature, without the observance by all or nearly all people of the "laws of nature," would essentially be non-existent (*Leviathan* Part I, chapters 13–4).

Hobbes's formulation, to this point, seems to allow plenty of room for free will alongside an important place for "law-governed behavior." In other words, in apparent contradiction to the postulate above that a philosopher must decide between free will and law, Hobbes seems to allow room for both, in a quite respectably rational position. One might conclude, up to this point in the argument, that humans, perceiving the evils of the war of nature, freely choose to embrace obedience to the laws of nature to ensure their own survival.

But that is not Hobbes's conclusion. Rather, he argued, humans obey the laws of nature because they have no other choice. Either one obeys, or one swiftly dies—there is no choice involved, except, of course, the choice to live or to die (*Leviathan* Part I, Chapter 6). Assuming one prefers to live, one is "forced" to obey the Laws of Nature. Now one might try to disagree with Hobbes on this point by arguing that even this formulation preserves a role for free will. For example, if a robber puts a gun to one's head and demands "Your money or your life," an example, by the way, that goes some distance toward recapitulating in a nutshell Hobbes's "state of nature," one always has the "choice," existentially speaking at least, to forfeit one's life rather than one's wallet. Free choice, though here reduced to a mere skeleton of its robust possibilities in a more safe and secure condition, still exists.

But Hobbes forecloses this possibility with another argument, one that Darwin apparently found quite attractive. The argument is that humans *always* choose with an eye toward what they believe will bring their overall greatest pleasure. In other words, the "choice" between living (and seeking pleasure) and dying is not perceived as a "choice" by most people, and certainly not by other non-human living organisms. Each potential decision presents the one who is deciding with a certain amount of anticipated pleasure. The rational decider weighs the various possibilities, and then chooses in accordance with where the greatest pleasure is thought to lie. In other words, no one chooses, or even can choose (within the caveat mentioned above), against the attraction of the greater overall pleasure.

Hobbes's argument has been criticized for, among other things, failing to discriminate between long-term and short-term pleasures, failing to weigh accurately the force of "mental" pleasures over "physical" ones, failing to notice the absurdity of choices for "pleasure" that actually result in pain, and failing to notice that some pleasures simply cannot be weighed on the same scale, i.e., that they are incommensurable (whereas Hobbes's theory seems to require that all pleasures are commensurable). But, whether such criticisms do any damage against Hobbes—and it may easily be doubted that they do—Darwin did not seem to notice them. His theory of choice, in the end, seems thoroughly Hobbesian. Humans "choose"—as do all animals—with an eye to securing the overall greatest pleasure in all their choices. This is not freedom; it is determinism.

One finds Darwin explicitly embracing some such position as this in the *Notebooks*. The first clue, if you will, that his thought was moving in this direction comes in *Notebook* M, in an almost off-hand throwaway comment about plants.

Plants, he reasoned, have no "sense" at all, and so lie outside the boundaries of any proper discussion of free will. But in their very lack of sense, they imply something important about life forms that *do* have sense, namely, that pleasure and pain have important roles to play in determining choice:

> With respect to free will, seeing a puppy playing cannot doubt that they have free will, if so all animals, then an oyster has & a polype (& a plant in some senses, perhaps, though from not having pain or pleasure actions unavoidable & only to be changed by habits). Now free will of oyster, one can fancy to be direct effect of organization, by the capacities its senses give it of pain or pleasure. If so free will is to mind, what chance is to matter (*CDN* M-72).

The first point that emerges from this passage is that free will in the organisms that possess it requires the organism having some sensation of pleasure or pain. Plants, by presumption lacking any such sensations, are outside the conversation about free will. Their behaviors are "unavoidable" precisely from lacking sensations of pleasure or pain. That observation points to the second and important point: what we call "free will" is tied in a basic sense to the *possession* of sensations of pleasure and pain. But here again we see the illusory nature of the phenomenon in question. Animals with the senses of pleasure and pain are really no more "free" than their floral counterparts. In the latter, actions are "unavoidable" for want of sensations of pleasure or pain. In the former, although sensations of pleasure and pain do exist, they lead to actions that are just as "unavoidable" as the behavior of plants. An oyster, to revert to that example, does what it does as a "direct effect" of its organization, where its organization is understood to be the cause of the particular sensations of pleasures and pains that it experiences, and where these sensations are understood to determine its behavior. From these observations Darwin again concluded, "free will is to mind, what chance is to matter" (*CDN* M-72). In short, both free will and chance are illusory.

THE "OLD AND USELESS NOTES"

But what about humans? Did Darwin see anything exceptional in humans' ability to reason, to weigh options, to resist some pleasures for the sake of a higher good, in other words, to exert free will and to avoid his deterministic conclusions about plants and non-human animals as nothing more than automatic pleasure-seekers? The answer is obscure because Darwin was reluctant to face the question head-on, except in his most private musings, those contained in what he called the "Old and Useless Notes." It should be said at the outset that Darwin's choice of a title for these notes suggests that it was one set of musings he did not wish to come before the eye of public scrutiny. And indeed, one has the sense reading these notes that one is confronting Darwin's most private thoughts, those thoughts where he pushed himself to disclose to himself the full limits of the implications of his theory and where he, at some point decided, here I must stop. The "Old and Useless Notes" (1838; hereafter "OUN") is in one sense the most exciting and revealing body of work in the corpus of Darwin's writings, public or private.

Right away Darwin saw that the crucial issue in choice for humans centers on the question of motivation. However it may be with plants and non-human animals, human beings do what they do on the basis of their motivations. Without a motive to do something, nothing interesting happens. (This statement must be qualified with respect to what Hobbes called "involuntary motions," e.g., the circulation of the blood, the inhalation and exhalation of the lungs, in general biological functions on which life itself depends (*Leviathan* Part I, Chapter 6). A "choice" is essentially an action undertaken on the basis of a prior thought or conception about a future desired state, a "motivation" in Hobbes's terms. If an agent believes a particular action will result in a state or condition that is anticipated to be more pleasurable overall than painful, the agent will "choose" that action.

> Every action whatever the effect of a motive –[must be so, analyse ones feelings when wagging one's finger—one feels it in passion, love—jealousy—as effect of bodily organisms—one knows it, when one wishes to do some action (as jump off a bridge to save another) & yet dare not—one *could* do it, but other motives prevent the action ("OUN" 25, emphasis in original).

Humans act from their motivations. Motives compete with one another. The motive to jump off a bridge to save another is a powerful motive, but even such a noble motive might be overridden by another motive (e.g., the desire to avoid harm (pain) to oneself). It is not always the case that the desire to avoid harm to oneself will prevail: some noble persons will actually find greater pleasure in the anticipation of saving another, and so will dive in. But for the one who chooses to dive in, as much as for the one who shuns duty for the sake of self-preservation, the active cause of behavior is the anticipated greater pleasure. Choice unconstrained by prior conceptions of anticipated pleasure or pain, which are themselves unchosen, is effectively blocked as a meaningful explanation for human decision-making.

EFFECTS OF EDUCATION AND NURTURE

One may wish to argue with Darwin on this point, as he realized. Is it not possible that the animal instincts humans possess might be overcome by education and nurture? Is it not possible that humans can be taught to be good, so that whatever they may desire in an unenlightened state might be improved and reformed by a good education?

Darwin anticipated this objection. Motives, he argued, are as much "given" to humans as is their physical constitution. But unlike physical constitution, motives are subject to change under the effect of environmental conditions, such as, for example, education.[4] Education can make a difference in how people choose. This may sound "Lamarckian," but it is not. The key factor in a Lamarckian explanation for evolution is heritability of acquired characteristics. Environments can and do change organic structures and behaviors. But if those acquired traits are not passed down, they are of no interest to evolutionary theory. To use a Darwinian example, the fact that one's parents have acquired a strong moral sensibility from studying the Bible (or any other moral text) has no predictive value for the morality that will exist in the offspring. The moral sense dies with the parents, so to speak.

Is this true of all environmental influences, including education? Darwin's view was uncompromising: one no more chooses one's education than one's parents. The important point is that after all the influences that shape how people think have played their roles, people are still governed by the motives that have been laid down in the basic mental structures of decision-making. Nature and nurture are both important, but neither has any way of overcoming the basic fact of human nature, that choices will always be toward what is perceived as the most pleasurable, however that perception has been brought into human consciousness. A great deal of how education influences choice must be assigned to chance:

> [Effects of hereditary constitution,—education under the influence of others.—varied capability of receiving impressions—*accidental* (so called like chance) circumstances. As man hearing Bible for first time, & great effect being produced.—the wax was soft,—the conditions of mind which leads to motion being inclined that way] one sees this law in man in somnambulism or insanity. Free will (as generally used) is not present, but he acts from motives, nearly as usual but by strong invariable passions—when these passions, weak, opposed and complicated, one calls them free will.—The chance of mechanical phenomena ("OUN" 25, emphasis in original).

The idea of weak passions giving rise to the impression of free will is drawn directly from David Hume. A person, not being strongly drawn in one direction or another, but having slight desires in both directions, or strong desires that are nearly equal in force, "fancies" that the choice is his, that he has free will to decide. But that is an illusion. The stronger passion will prevail in every case, even if it is only slightly stronger. A page later Darwin continued:

> It may be urged how often one try to persuade person to change line of conduct, as being better and making him happier.—He agrees & yet does not.—because motive power not in proper state.—When the admonition succeeds who does not recognize an accidental spark falling on prepared materials. From *contingencies* a man's character may change—because motive power changes with organization. The general delusion about free will is obvious—because man has power of action, & he can seldom analyse his motives (originally mostly INSTINCTIVE, and therefore now great effort of reason to discover them: this is important explanation) he thinks they have none ("OUN" 26, emphasis in original).

As these passages suggest, Darwin was particularly impressed with the role of contingency or accident in determining what education a person receives. This parallels the role of "chance" in altering a plant or animal's organization in life's evolutionary process. A "man hearing Bible for first time" may be greatly affected in his motivational scheme of preferences. He may "change" his ideas from those he had acquired through heredity, instinct, and prior training. But the entrance of the Bible's teachings into a person's life is no more a chosen event than are his original inclinations. This idea is elaborated in a pair of footnotes appended by Darwin to page 26 of the "OUN":

> A man may put himself in the way of Contingencies.—but his desire to do so arises from motives.—& his knowledge that it is good for him the effect of Education and mental capabilities.—Animals do attack the weak and sickly as we do the wicked.—we ought to pity & assist & educate by putting contingencies in the way to aid motive power ("OUN" 26).

In other words, even a person's desire to "put himself in the way of contingencies," so as to enlarge his range of choices through better understanding and knowledge, is itself merely a consequence of prior, unchosen, motives. Darwin thus did not dispute that humans make choices, even choices to improve themselves. What he disputed is that such choices to do so are in any sense "free" of prior motivational impulses; and also, that any new motivational impulses that may arise as a result of new learning and education are any less constrained by the necessity to choose the greater over the lesser pleasure than was the original motive to seek improvement in the first place. In a marginal notation to his copy of John Abercrombie's

Inquiries Concerning the Intellectual Powers and the Investigation of Truth (1838), Darwin commented:

> Yes, but what determines his [a person's] *consideration* [to make a particular choice to seek education]—his own previous conduct?—and what has determined that? And so on.—Hereditary character and education—and *chance* (aspect of his will) circumstances ("OUN" 25, n. 25-1, emphasis in original).

Darwin was quick to draw two important conclusions from this line of reasoning. The first concerned implications of his view for a proper conception of human morality, a subject treated in the next chapter. The second was his realization that, by dismissing free will as a meaningful concept for understanding human behavior, he had effectively assimilated human to non-human behavior among all biological organisms. Everyone agrees that plants have no free will, and most agree that non-human animals have no free will. Yet both groups of organisms display life and motion. If humans have no free will, therefore, it seems to follow that whatever explanations account for plant and animal behavior will account for human behavior as well. Darwin cast this idea in terms of "two great systems of laws," one governing the inorganic world, the other the organic. A separate "human" system is not mentioned. Laws of the organic world apparently have a direct connection to laws governing inanimate matter.

> Hence there are two great systems of laws <<in the world>>, the organic and the inorganic. –The inorganic are probably one principle for connect[ion] of electricity [and] chemical attraction, heat and gravity is probable.—And the Organic laws probably have some unknown relation to them.—In the simplest forms of living beings namely <<*one individual*>> vegetables, the vital laws act definitely (as chemical laws) as long as certain contingencies are present (contingencies as heat, light, etc.). This is true as long as movement of sensitive plant can be shewn to be direct physical effect of touch & not irritability, which at least shows local will, though perhaps not conscious sensation. – During growth [laws?] unite matter into certain form; [this is] invariable as long as not modified by external accidents, & in such cases modifications bear fixed relation to such accidents ("OUN" 34 and marginal notes, emphasis in original).

Biological organisms are governed by "fixed laws," animals as well as plants, and by extension, humans. These laws operate in definite (i.e., predictable) ways, assuming the presence of the proper external conditions, or "contingencies." But even when these contingencies change, the resulting change in growth is directly tied to the new contingencies. Nurture can alter growth, but it cannot supply an organism with an ability to overcome the operation of laws. One must assume Darwin meant that organisms can and do *respond* to altered conditions, but the response is as fully a result of "laws" as the organism's behavior in unaltered conditions.

Plants and animals no doubt differ from each other in one important respect: the existence of "consciousness" among the latter but not the former:

> In animals, growth of body precisely same as in plants, but as animals bear relation to less simple bodies, and more extended space, such powers of relation required to be extended. Hence a sensorium [in animals], which receives communication from without, & gives wondrous power of willing. These *willings* are common to every animal [and they are] instinctive and unavoidable.—*Can the word willing be used without consciousness, for it is not evident, what animals have consciousness*. These willings have relation to external contingencies, as much as growth of tissue, and are subject to accident ("OUN" 35, emphasis in original).

But, as the passage makes clear, "willing" has no implication of "free willing." The "willing" is a function of a "sensorium," by which Darwin apparently means a seat of perception and sensation present in animals (including humans) and not plants.[5] Sensorium is connected to "extended relation" (i.e., motility) and "consciousness," two notions associated with animals but not plants. One might be tempted to think that by allowing consciousness and willing, Darwin was admitting free will. Plants react "automatically," animals and humans "choose." But that interpretation misses the mark:

> Sexual willing [in animals] comes on [during certain] period[s] of year as much as inflorescence [in plants]. Why [should animals enjoying their movements show] more [evidence of willing] than movement of sap or sunflower to sun? I should think there was direct <<physical>> effects of more or less turgid vessels; effect of heat, light, or shade ("OUN" 35).

And, to clinch the case that man is not special with regard to willing, Darwin added:

> It is easy to conceive such movements & choice, & obedience to certain stimulants without conscience [consciousness?] in the lower animals, as in stomach, intestines, & heart of man. How near in structure is the ganglionic system of lower animals & sympathetic of man ("OUN" 35 and marginal notes).

How is "willing," then, connected to the role of "chance" or "accident" in Darwin's account of biological behavior? In the "OUN" he remained committed to the idea that change in organisms is governed by "laws," although in these Notes he refrained from speculating about what these laws might be. But laws do not account for everything. All organisms, he now argued, are exposed to "contingencies," that is, physical conditions that are themselves not the result of known laws at all but are purely "by accident," as far as we can tell. Laws no doubt govern everything in nature, but like the outcome of a toss of dice they may be of little predictive value. Physical conditions may work differently on different organisms, but one fact about them seems certain: they are not *essentially* different in their role in effecting change in animals than in plants, or indeed in humans than in non-human organisms.

It may seem to human contemplation that one important contingency, a person's education, and by extension consciousness, produces a qualitatively different sort of effect, enabling humans to choose among their appetites and desires. But that is a misconception. For one thing, education itself is no more the result of a free act of choosing than other environmental contingencies. It is true that humans may "choose" to pursue a formal education; but on closer inspection that "choice" is just as constrained, and necessitated, either by the interposition of other people in causing the education to come about or by a prior desire within the agent himself to seek out improvement, as are other, more straightforward choices for pleasure.

Secondly, once the effects of education have taken root in human consciousness, a new set of desires arises to replace or supplement an earlier one that had come about through heredity and early upbringing. These new desires are not freely chosen, any more than the education was. They depend no doubt on a complex interplay between laws of human psychology, the nature of the education, and a person's time and place on this planet. But however that question resolves itself—and Darwin attempted no answer to it—the resulting product, a new human consciousness, is as fully bound as the old one was in its preferences and decisions by the new set of appetites and desires that have been brought into being. Of this Darwin exhibits no doubt.

NOTES

1. The issue of "free will" in Darwin's thought has not been the focus of much scholarly attention. An exception is R. J. Richards (1989, 122–4; 240–1). The interplay between "free will" and "determinism" as a philosophical dilemma (which it was for Darwin) is examined in Gigerenzer et al. (1989, ch. 8).

2. An objection might here be raised: just because someone is hungry does not mean that she will eat, even if palatable food is available. She may "choose" to walk away from the object of desire. Does that possibility not demonstrate the existence of a "free" will? Hobbes, and more

explicitly Hume, answered this objection. The objection only begs a question: why did she walk away? Presumably because of a stronger (yet ultimately unchosen) desire pulling in a different direction—the desire to stick to a diet, the desire to be fit for the upcoming race, or any number of other possibilities. The point here is only that the desires are unchosen; how one *acts* in the face of given desires depends only on what is possible and what the pecking order of competing desires happens—chances—to be. Hume goes further: what if, even in the face of *all* competing desires, one still "chooses" to ignore all of them, just to "prove" free will exists? If it is not obvious already, the desire to "prove" free will exists is, well, just another (unchosen) desire.

3. The argument may be seen to come down to a basic proposition about so-called "free will"; either oysters have it too, or else humans do not. Darwin appears not to have been too concerned with which way to express the idea. In the *Notebook* entries considered here, he was inclined to say that no organisms possess free will; between an oyster and a human no sharp line can be drawn. But recall that in correspondence with Charles Lyell (*CCD* vol. 8, 262 [June 1860]) he took the opposite position, when considering dinosaurs. Lyell wanted to maintain, against Darwin's opinion that humans are "just another step" in the evolutionary sequence, that humans are unique by possessing free will—something not present among the dinosaurs. Darwin countered with the view that dinosaurs had free will as much as humans do. This is not a contradiction. It merely shows that Darwin saw the "free will/determinism" distinction (to separate humans from the rest of creation) as a red herring. One may say dinosaurs have free will or that humans do not have it: it comes to the same thing in either case. Additional discussion in chapter 3 of the present work.

4. This statement must be qualified by the acknowledgment that Darwin always did allow a role—always a small one—for Geoffroyian and Lamarckian possibilities that the environment could act to transform the physical organization of organisms. Usually, however, such changes came about "by chance." See chapter 8 of the present work for a discussion of the issue.

5. "Sensorium" is a term Darwin redeployed much later, in the *Expression of Emotions in Plants and Animals* (1871). It almost disappears from his writings in the interim between 1838 and 1871, giving rise to the speculation that when he brought it back into play in the *Expression* he was copying from notes that he had recorded 35 years earlier.

10

Chance and Human Morality

FROM THE FOREGOING we might want to conclude that Darwin could not possibly claim to have a moral philosophy. Saddled as he was with "chance" and "determinism" as the only available accounts of why and how humans are cognitively organized as they are and act as they do, no room, it seems, is left for the operation of moral character or moral agency.

That conclusion must be forestalled. Before we begin, however, we should draw a distinction between Darwin's own "moral code," and what we may call his "moral philosophy." Anyone who spends much time with Darwin's writings, especially his private correspondence, will find nothing but a "decent" and "morally respectable" person as those qualities were defined in nineteenth-century Britain. Recent biographies have treated this subject at sufficient length to dispense with the need for more detailed treatment.[1]

But did Darwin have a "moral philosophy," that is, a systematic and integrated set of beliefs that would or should help someone holding it come to morally correct decisions? Again we must draw a distinction (following Richards, 1989, 113–20, and other authors he cites) between an "ethical philosophy," in which normative claims about proper moral behavior are made, and an "anthropological" or "scientific" account, in which the philosopher-naturalist tries only to give a rational and coherent account of how moral beliefs may have arisen in human communities in the first place and why they may have been "selected" for survival.[2] The second path, the one Darwin took, did not commit him to endorsing or rejecting any particular moral scheme. His job, as he saw it, was purely analytical: why do some moral schemes survive and others fail? We are not surprised that his conclusions provided additional support for his favorite and most important idea: natural selection. In a nutshell, Darwin found that the prevailing moral schemes of his day (more than one existed) or of any day could all be resolved through analysis into an explanation that could show *these* moral schemes rather than others would be "selected" because they favored for survival the peoples who were first to seize on them.

DESCENT OF MAN 1872

I will focus most of the discussion that follows on the early *Notebooks* because that is where Darwin first started to sort out his ideas about human morality, mainly between 1838 and 1840 (Barrett, in Gruber, 1981). But that one-sided emphasis may seem distorting. Darwin notoriously avoided the "human" question in the *Origin*, all editions, except to say his theory might "shed light" on human evolution (including mental evolution). But he did engage fully with human evolution in his later book *Descent of Man* (1871). Did he change his mind between 1838 and 1871? If we are to know what the mature Darwin believed about human morality—what he believed about moral theories and how human morality is to be explained—we cannot overlook his later reflections.

The discussion of this issue will be brief. The main reason is that almost all of Darwin's reflections about human morality in *Descent* trace a direct ancestry back to his 1837–1840 *Notebooks* and notes. Indeed, one gets the impression that nearly everything he had to say substantively in *Descent* was recycled material from the *Notebooks* and early notes. I will only note the key features, providing textual citations as needed.

(1) "Gradualism": the idea, borrowed from Charles Lyell's great *Principles of Geology*, is that "change" in the shape and structure of the earth and its features is slow and gradual, but inexorable. Lyell did not apply this idea to organic structures, but Darwin did, and that was a major advance in establishing Darwin's theory as a correct account of the modification of species. An important claim in the *Descent* is that "gradualism" applies not only to organic structures but also to mental phenomena. Against the tradition of thought that claimed "man" was *qualitatively* distinct from all other creations by virtue of possessing "reason" and "free will," Darwin saw only small quantitative steps from "lower" to "higher" with no clear line of demarcation:

> My object in this chapter is solely to show that there is no fundamental difference between man and the higher mammals in their mental faculties. Differences... between the highest men of the highest races and the lowest savages, are connected by the finest gradations (*Descent*, vol. 1, ch. 2, 34).

The mental faculties of "man" are no doubt more complex than those of other animals but essentially no different.

(2) "Free will": If there is any difference between the early *Notebooks* and notes and the *Descent of Man* as regards free will, it is only that by the time of the *Descent* Darwin had simply dropped the subject. He had, in fact, already stopped talking about free will much earlier. It barely shows up in works he wrote after 1842, with one interesting exception. In the early 1860s, when Charles Lyell was objecting to Darwin's "stone-house" metaphor because it did not allow any room for human free will, Darwin disagreed, but for a strange reason. Lyell had claimed that free will only came along with "man," and that organic nature up until man appeared did not have it or need it. For example, Lyell noted, the dinosaurs did not have or need free will. Darwin's view was different: not that "man" does

not have free will, but that dinosaurs *did* have it! One can only surmise that Darwin was thinking in typical gradualist terms, meaning that he saw no sharp line of separation between lower animals and man as regards mental capacities (see paragraph one above in the present chapter on "Gradualism"; and chapter 6 of the present work for a full discussion).

We may wonder why Darwin avoided talking about free will after 1842. The answer is one he gave himself in private messages to Charles Lyell and Asa Gray in 1861: the dispute between "free will" and "determinism" is a "wretched embroglio" (*CCD* vol. 9 [1 August 1861], 225, to Lyell), and as a general rule Darwin wished to stay out of such "thick mud." Asking him to answer questions about "design" and "free will" would be about as fruitful as asking a gorilla to make out the first book of Euclid's Geometry (*CCD* vol. 9 [December 11, 1861] to Asa Gray). Darwin's typical answer was, "no thank you."

(3) "The Moral Instinct": This subject will be developed in greater detail directly. That discussion will be based mainly on the early *Notebooks* and notes because that is where Darwin first worked out his views about ethical theory. The only important point here is that again I discern no real change in Darwin's moral theory as presented in the *Descent* in 1871 from what he had already concluded by 1840. Prominent among the ideas in his thoughts on this subject are: (1) that humans possess an innate "moral instinct"; (2) that this instinct may be modified though experience and education, and it generally needs to act in support of the historical-cultural conditions in which it operates (hence the element of "moral relativism" in Darwin's thought); (3) that despite differences from one culture to another in moral codes and beliefs, all moral schemes do seem to possess a common, or universal element: choosing characters and behaviors on the basis of their contribution to the good of "the group" or "community"; these are the moral aspects that are most likely to be "selected"; (4) moral choices are motivated by a desire for pleasure. What ensures that choices for morally correct character and behavior will be selected over choices for more fleeting but less worthy objects is the greater *durability* of the pleasures achieved through the former.

(4) "Variation": Perhaps the most striking aspect of continuity between the early and the late Darwin are his thoughts on variation. He struggled with a proper account from 1837 all the way through the *Variation of Animals and Plants Under Domestication* (1868). But really very little ever changed. The subject is taken up, yet again, in Chapter IV of the *Descent*.

What we find in this chapter is: (1) a new title: the words "Causes/Laws of Variation" have disappeared, replaced now by the expression "On the Manner of Development of Man from Some Lower Forms" (103); (2) commingling of "laws" with "causes" of variation, as in earlier works (see chapter 3 of the present work); (3) enumeration of the causes of variation, including the now-familiar "direct action of conditions of life" (i.e., Geoffroyism); "effects of the increased use and disuse of parts" (i.e., Lamarckism); "arrested development"; "reversion"; and "correlated variation." Darwin then gives examples of each of these "causes/laws" (109–26).

What is new here? Actually, nothing, except for the significant *omission* of "chance" as a cause! Darwin had only one word to say about this, using the expression "spontaneous variation" rather than chance, and this short statement is entirely overshadowed by the extensive treatments given to other causes:

> Besides variations which can be grouped with more or less probability under the forgoing heads, there is a large class of variations which may provisionally be called spontaneous, for they appear, owing to our ignorance, to arise without any exciting cause" (126).

Did Darwin change his mind about "chance" in the production of nature's variations? I doubt that he did. The *Descent* is merely further evidence that he was more determined than ever to advance his campaign against the *words* "chance and accident" without giving up on the idea.

FREE WILL AND MORALITY

At first it may seem impossible to believe that any moral theory that excludes a role for free will should be considered a *moral* philosophy. Take away "free will" and not much of the structure for any modern moral philosophy is left standing. We do not accuse termites of being "evil" even if what they do to our homes is destructive. They are programmed to survive by eating wood. If humans are essentially no different—doing what they do because they too are programmed—where is the room for morality, either in theory or practice?

But Darwin did not come to the conclusion that no room for human morality exists within the confines of his theory. He did not say that there is *nothing* special about "man." Humans, unlike all other animals, not to mention plants, do display "moral" behavior, and that is a "good thing" from the standpoint of human survival and thriving. This brings us to the second important conclusion Darwin derived from his speculations about free will. Humans differ from all other life forms in two important respects, neither of which preserves a coherent notion of free will within his theory, but both of which are worthy of notice for *confirming* the determinism that has now come to pervade Darwin's views about human life.

The first is that humans' ability to use reason enables them to transform and appreciate their environment in ways that are simply unavailable to other life forms. They can and do transform the very character of their surroundings, improving and complicating life in remarkable ways, just because they can see further and plan better than all other creatures. They may even create wondrous productions that "thrill" the senses. In this sense humans seem to be in control of their own destinies in a way that other creatures are not:

> I grant the thrill, which runs through every fibre, when one behold[s] the last rays of [sun] && or grand chorus are utterly inexplicable. I cannot think [this observation provides] reason sufficient to give up my theory. Viewing from eminence, the wide expanse of country, netted with edges & crowded with towns & thoroughfares, I grant that man, from the effects of hereditary knowledge has produced almost greater change in the polity of Nature than any other animal ("OUN" 7).[3]

What is to be noted in this passage—and others could be cited—is that humans experience and do remarkable things that are unique to them. At the same time, however, these events and experiences do not overthrow the main argument of the theory. These wonderful effects are as much a result of heredity and (acquired) knowledge as are the productions and experiences of all other creatures. The presence of "reason" in human actions and experiences does not make a dent in the overriding view that humans are somehow different from animals and plants from the standpoint of free will. One might say that humans have stronger powers of appreciation and greater abilities to change their environment than other creatures. But that is far different from saying they have greater powers to choose freely what they experience or do, or freely to choose morally correct behavior.

The other major conclusion Darwin derived from his moral speculations concerns how moral conduct among humans is properly to be understood. Competing with Hobbes's idea that all moral behavior may be resolved into an account in which moral choices are determined by rational agents weighing whether a particular moral action is more or less likely to result in personal survival (the account from "rational self-interest," as we shall call it) was an idea of a "moral sense," popular in Darwin's day from its promotion by a group of Scottish and British philosophers whom Darwin had read. To abbreviate, this school of thought held that humans have an innate moral conscience, inbred from birth, an inborn moral compass, if you will. This is an idea that hearkens back to the New Testament theology of St. Paul. The theory states that a "law of goodness" is inscribed in every human heart. Goodness consists in nothing more than consulting and obeying that law. Evil is thus to be explained by some people's failure to read what is written in their own souls or to heed what they have read. Goodness, by contrast, is a matter of reading the law and obeying it. The important point for us is that the moral code, in the Pauline version, is inscribed in human nature itself. It does not need to be learned through additional instruction, although instruction may be useful to make the message (already present) fully clear to those who already possess it.

Darwin wrestled with these ideas in the "Old and Useless Notes," not in their Christian exposition as they appeared in the New Testament, but in their appearance in his own time in the writings of other philosophers. He understood the question of morality to be one of deep controversy in his day, suspended between two great schools of thought, that of "Kant and Coleridge," on one side, and that of "Locke, Bentham, and Hartley" on the other ("OUN" 33; recent discussion in Dilley [2013]). This is not the place to enter into the particulars of the dispute between these two schools. Suffice it to note that, for Darwin, the central issue came down to the question whether knowledge in general, and moral knowledge in particular, is "instinctive" (or, as we might say, innate), the Kantian position, or acquired by experience (the Lockean view). Darwin perhaps was not sure, but he seems to have leaned toward the former:

> Is this not almost a question of whether we [humans] have any instincts, or rather the amount of our instincts—surely in animals according to [the] usual definition, there is much knowledge without experience. So there *may* be in men ("OUN" 33).

This passage, though, revealing as it is about Darwin's opinion about "innate knowledge," does not advance us in understanding his opinion about innate *moral* knowledge. Does this exist? The answer he gave is, yes and no. A "moral taste" does exist "in us," like other hereditary instincts. But it would be a mistake to conclude that this "moral taste" is anything other than an inborn preference for some states of affairs rather than others that have previously served survival needs. *Post facto*, humans come to call these preferred states of affairs "good," but in truth they are no more "morally good" than pleasant sensations of taste in one's mouth are morally good:

> Our tastes in mouth by my theory are due to hereditary habit (& modified and associated during lifetime). So is our moral taste.... [Mackintosh makes] some good remarks on analogy of pleasure of imagination <<the utility part

> being blended and lost>> and moral sense. My theory explains both, perhaps by habit ("OUN" 50v).[4]

Darwin, in other words, did not doubt or dispute the existence in humans of what is called by some philosophers a "moral sense." What he disputed is that this "sense" is any different qualitatively from natural instincts in animals, proclivities inborn that draw them toward certain objects and steer them away from others. In these cases one may agree the instinct is "implanted" by nature, yet disagree that it has any claim to a "moral" status, that is, to a status deriving from the chosen action being contrary to the mere pull of pleasure because the agent considers acting against the mere pull of pleasure to be the morally proper thing to do. The chronological order for Darwin is the reverse. Humans are drawn to certain behaviors because of the pleasure, then afterward decide that these actions are rightly called "good."

> It appears that Sir J. [Mackintosh] & others think there is a distinct faculty, of conscience.—I believe that certain feelings & actions are implanted in us, & that doing them gives pleasure & being prevented [gives us] uneasiness, & that this is the feeling of right and wrong.—So far it has *independent* existence & is supreme, because it is <<a>> part of our nature, which regulates our feelings steadily & not like our appetites & passion, which receive enjoyment from [immediate] gratification & hence are forgotten—only so far do I admit its *supremacy* ("OUN" 52, emphasis in original).

"Supremacy," in other words, is to be accorded to the "faculty of conscience" in humans, *not* because it somehow overrides the pull of immediate pleasures, but rather because the pleasures it identifies as worth pursuing are more steady and long-lasting. It is their durability, not their greater moral worth, that makes them "supreme" over other appetites.

> Butler & Mackintosh characterize the moral sense, by its "supremacy"; I make its supremacy, solely due to greater duration of impression of social instincts, than other passions, or instinct.—Is this good? ("OUN" 54).

This theory, Darwin believed, was far superior to any theory that posited a qualitatively superior moral sense in humans:

> NB. [Both Hartley and Mackintosh argue that] conscience checks the *wish* to outward gratification, whilst no desire of gratification will check the conscience's desire for virtue. [I expect there is some fallacy here:][cd]. [They believe]

> the very end of conscience is [to] stop [the] wishes of passion &c. whilst the passions have no relation [to conscience?]. I think this <boshes> <<nonsense>>. My theory of durableness will explain it ("OUN" 54v, emphasis in original).

In other words, what is often called conscience (i.e., a desire for virtue) is only another passion for gratification, like all others. It does not, and cannot, *check* the other passions, only outlast them. The other passions do not "stop" being passions or to have "wishes"; they only yield to the greater durability of the desire (i.e., the feeling of pleasure that "acting virtuously" brings). Darwin sums up as follows:

> Perhaps my theory of greater permanence of social instincts explains the feeling of right & wrong, arrived at first <rationally> by feeling—[then] reasoned on, steps [of reasoning] forgotten, habit formed, & such habits carried on to other feelings, such as temperance, acquired by education. In similar manner our *desires* become fixed to ambition, money, books, &c &c ("OUN" 55, emphasis in original).

This was Darwin's last entry into this most fascinating of his notebooks. He ended much as he began, a determined determinist who saw, or at least believed, that what was taken as moral virtue in his day was no different in essential respects from any other passion or desire. All passions, including the passion for virtue, are inborn and innate. All passions in humans, and perhaps in some species of non-human animals, may be altered and redirected by education and nurture, but such external influences do not and cannot alter the fundamental drive among all creatures to pursue what appears to be the overall greater pleasure. If the passion for virtue among humans is different at all from all other passions among the organisms inhabiting the animal kingdom it is only in regard to its greater steadiness and durability.[5] Most everyday passions and desires come and go. They are either satisfied or not, but one can be sure that, either way, tomorrow will present a new set of desires different in some ways from yesterday's. This is not so in the case of the desire to be virtuous: it persists over long stretches of time, ensuring that most people most of the time will find their greatest satisfactions in being "good" and in doing "good things."

PUNISHMENT AND REWARD

It is, of course, quite a different question whether there is anything truly praiseworthy in conduct that is chosen only for the sake of an unavoidable desire for anticipated pleasure. We may dislike termites for gnawing away the foundations of our homes, and we may like our domesticated pets for being friendly and cute. But we normally do not believe their actions are morally contemptible or praiseworthy. We may, it is true, exterminate termites and throw our pets a bone, but it seems contrary to our basic intuitions that we are thereby "punishing" or "rewarding," as though the termites and pets *deserved* these things or had *earned* them, or *merited* them. On the contrary, we believe the termites and pets cannot help but do what they do. It is just that, in the case of the termites, what they cannot help doing turns out to be fatal for them when the exterminator comes, and in the case of the pets, what they are incapable of avoiding, friendliness and cuteness, makes them the lucky beneficiaries of having happened to belong to these particular owners who gives bones to dogs they happen to adore.

Our question for Darwin is, do we find anything substantially different in the human realm? Are humans, like termites and pets, mere automatons who cannot help what they do, and so are below the possibility of moral goodness? Darwin's answer is complex. Certainly, he argued, "wicked" people cannot be "blamed" for their wickedness, any more than sick people can be "blamed" for their sickness. Both states are a consequence of powers beyond an individual's control—his innate instincts and education, in the first case, his physical constitution and environment in the second:

> One must view a wicked man, like a sickly one. We cannot help loathing a diseased offensive object, so we view wickedness [in the same way]. It would however be more proper to pity than to hate & be disgusted with them ("OUN" 26).

A page later:

> This view should teach one profound humility; one deserves no credit for anything (yet one takes it for beauty and good temper), nor ought one to blame others ("OUN" 27).

A person deserves no more "credit" for being morally good or bad than for being beautiful. Yet, of course, people do claim credit for moral qualities and conduct, and beyond this, for having qualities of character like a "good temper." Darwin affirms rather than disputes that people in fact do claim credit and assess blame for how people act and how people are. He only disputes that such claims and assignations are justifiable, not just in particular cases, but also *as such*.

This may seem a dangerous doctrine, as Darwin immediately realized. If there is no right and wrong, no good and evil, providing definite targets for deliberate human choice to aim at or avoid, it would seem, *everything* is permitted (the slogan, incidentally, of the Society of Assassins that Nietzsche praised in his *Genealogy of Morals* (1956 [1887], 287). One cannot rule out some behaviors, such as murder and rape, on the grounds that they are wrong, because no "wrong" exists. Nor is it possible, under this theory, to encourage some behaviors, like donating money to charity or helping the elderly, because no "right" exists. This is a moral nightmare, and Darwin knew he would have to deal with it.

Darwin appears to have fashioned a three-pronged assault on the moral anarchy he had just unleashed. The first prong was to "bury" these notes. He labeled them the "Old and Useless Notes" for the evident reason that he did not want anyone to see them or take them seriously. I am a little surprised, in fact, that he did not destroy these "Notes" altogether, but one can push a psychological explanation only so far. Whether he ever wanted them published I do not know, but he did not destroy them.

His second line of defense against radical moral skepticism was to reassure his readers (i.e., himself) that the doctrine, although true, would probably not do any damage in the outside world, even if, like a deadly virus, it were somehow to be released into the atmosphere. For one thing, few people would ever be likely to read these "Notes," and even if they did, only those who would read them with the greatest care would ever fully comprehend the new doctrine.[6] Most people are not easily swayed from their moral prejudices, and the few who are would be unlikely to make much of a public splash.

> This view [that neither credit nor blame can rightly be assigned to human choices] will not do harm, because no one can be really *fully* convinced of its truth, except man who has thought very much, & he will know his happiness lays in doing good & being perfect, & therefore will not be tempted, from knowing everything he does is independent of himself [i.e., his own free will] to do harm ("OUN" 27, emphasis in the original).

This may seem an odd statement in light of what has gone before, but it is not. Darwin knew certain behaviors to be "better" than others in one particular sense—the sense of social utility, a doctrine generally known under the word "utilitarianism." In some ways utilitarianism is an unusual candidate for a "moral" doctrine, because it makes no claim to be warranted on the basis of absolute values or a divine presence that oversees, judges, and requires morally correct behaviors. Darwin had no room for a transcendent morality of this kind:

> These views [i.e., my views] directly opposed & inexplicable if we suppose that the sins of man are under his control, & that future life is a reward or retribution. It may be a consequence but nothing further ("OUN" 28).

But Darwin did have room for a *social* morality, such as utilitarianism claims to have. Utilitarianism is often defined as the moral scheme that promotes by deliberate human design "the greatest happiness of the greatest number of people." Darwin's embrace of this sort of moral scheme was his third assault against the moral anarchy that was intimated by his theory. Although he nowhere developed a moral theory systematically, if he had done so it would have gone something like this: Humans, like all animals, are endowed with instincts and impulses that draw them by necessity toward objects of anticipated pleasure and repel them by like necessity from objects of anticipated pain. These native instincts and impulses may be modified by nurture and education, but such modifications only alter the *objects* of desire and aversion, not the fact that humans are drawn to these new objects by the same necessity that compelled them to pursue earlier desires. In other words, one searches in vain to find any place for a truly *free* will in any human behavior in Darwin's theory.

A "SOCIAL" INSTINCT

Yet, among the desires and aversions that determine human conduct is one that is especially useful in assisting humankind to survive and even flourish in a difficult world: a *social* instinct. This instinct causes humans to be concerned, at least to some extent, with the well-being of other people beyond self alone. This instinct may be defective in some individuals, but when it is it may often be improved and strengthened by education. "Believer in these views," he wrote, "will pay great attention to Education" ("OUN" 27). Recognizing this particular instinct commits the scientist to nothing more than recognition of a fact. But it permits the scientist to regard this particular fact as of special importance for human survival and flourishing. And if the scientist permits herself to allow that human survival and flourishing are "good," she may be allowed to affirm the moral worth of the social instinct, insofar as it furthers that good.

Darwin in fact did affirm that human life and flourishing are "good," and so did affirm the moral worth of the social instinct:

> Two classes of moralists: one says our rule of life is what *will* produce the greatest happiness.—The other says we have a moral sense.—But my view unites both <<& shows them to be almost identical>>. + What *has* produced the greatest good <<or rather what was necessary for good at all>> *is* the <<instinctive>> moral sense.... Society could not go on except for the moral sense, any more than a hive of Bees without their instinct ("OUN" 30, emphasis in original).[7]

Darwin also understood that the moral sense can, indeed must, change from one time to the next. What "works" for primitive peoples in holding society together may become useless or even counterproductive for more advanced peoples:

> These bad feelings [such as anger and shame] no doubt originally necessary [because in earlier times] revenge was justice. No checks were [at that time] necessary to the vice of intemperance, circumstances made the check.... Civilization is now altering these instinctive passions, which being unnecessary we call vicious ("OUN" 29–29v).

Indeed, the entire project of "civilization" may be regarded as a progressive alteration and taming of the ruder and stronger passions as they become less and less necessary to ensure the social peace and the larger social good:

> The origin of the social instincts <<in man and animals>> must be separately considered.—The difference between civilized man & savage, is that former is endeavoring to change that part of the moral sense which experience (education and the experience of others) shows does not tend to the greatest good.—Therefore rule of happiness is to certain degree right. The change of our moral sense is strictly analogous to change of instinct of animals ("OUN" 30v).

To this point Darwin may appear to be defending a version of moral relativism, according to which there are no "absolute" values, only values relative to particular times and places. Such a doctrine would thus leave open moral possibilities that would be possible only in a godless universe and that may strike a contemporary reader as horrendous and unacceptable. Darwin in fact did encounter at about the same time he was composing his "Old and Useless Notes" a version of moral relativism in Harriet Martineau's *How to Observe Morals and Manners* (London, 1838).[8] About her work Darwin remarked that she "argues <<with examples>> very justly [that] there is no universal moral sense" (*CDN* M-75).

Darwin at first tended to agree: "Must grant," he wrote, "that the conscience varies in different races," indeed, that such differences are "no more wonderful than [that] dogs should have different instincts" (M-75). Among the "examples" cited by Martineau was the fact that those men once most highly esteemed were those who killed the greatest number of enemies in battle (whereas today a more peaceful attitude is esteemed). Martineau also drew attention to the esteem once accorded to those Polynesian women who destroyed their young. From such examples she concluded, "Every man's feeling of right and wrong, instead of being born with him, grow up in him from the influences to which he is subjected" (Martineau 1838, 22, cited in *CDN* M-75, n. 1). That is a tidy summary of the essential principle involved in this conception of moral relativism.

But Darwin quickly came to modify a strictly relativist moral position by acknowledging some absolutes—as indeed, he came to see, Martineau had also done. Basic to his modified position is that some universally acknowledged ideas of right and wrong *do* exist:

> Mart[ineau] allows *some* universal feelings of right and wrong <<& therefore in *fact* only *limits* moral sense>> which she seems to think <<are>> to make others happy & wrong to injure them without temptation.—This probably is natural consequence of man, like deer &c, being social animal, & this conscience or instinct may be most firmly fixed, but it will not prevent others from being engrafted (*CDN* M-76–7, emphasis in original).

Can anything more be said about these moral "universals" than what we have here? For Darwin the answer again came down to the basic "fact" of a "moral instinct," an innate feeling implanted in animals, including humans, that makes them feel pleasure from acting well toward others and uneasy from acting poorly. David Hume had employed the idea of "sympathy" to describe this feeling, but Darwin's choice of term came instead from Martineau. It was "charity":

> Miss Martineau (How to Observe p. 213) says charity is found everywhere (is it not present with all associated [viz., social] animals?) I doubted it in Fuegians, till I remembered Bynoes story of the women.[9] The Chillingham cattle (& porpoises) have not charity [but humans do?] (*CDN* M-142).

Darwin then endeavored to connect this moral sense of charity with a more general account of the origin of human morality as a whole. He saw continuities between non-human animals and humans in terms of value systems and moral codes, but was of the opinion that humans' greater intelligence had produced more complex value systems than other animals were capable of. Nevertheless, all moral systems at any level were merely the sum of innate instincts, augmented in the case of humans, by additional moral instruction.

> May not moral sense arise from our enlarged capacities <<yet being obscurely guided>> or strong instinctive sexual, parental & social instincts, giving rise [to] "do unto others as yourself," "love thy neighbor as thyself." Analyse this out, bearing in mind new relations from language.—The social instincts

> [are] more than mere love.—fear [of] others acting in unison. Active assistance, &c &c. It comes to Miss Martineau's one principle of charity (*CDN* M-150–51).

The role of reason and memory in enlarging the moral sense among humans is acknowledged again in the "OUN":

> As man has so very few (in adult life) instincts.—This loss is compensated by vast power of memory, reason, & & many general instincts, as love of virtue, of association, parental affection.—The very existence of man requires these instincts…Conscience is one of these instinctive feelings ("OUN" 37 RHC (right hand column)).

Darwin's moral theory, then, if it is permissible to speak of such a thing in light of his disjointed musings in the M, N, and "OUN" *Notebooks*, comes down to this. Humans are indeed moral creatures, not merely in the sense that they have distinct ideas of right and wrong, but in the deeper sense that such ideas are grounded ultimately in a "moral sense," a "social instinct." This moral sense is, it is true, capable of modification over time, giving rise to the suspicion that Darwin was a moral relativist, and so, as this implies, an atheist. But Darwin was never comfortable with that position, or that label. Instead he found a satisfactory account of morality in the idea that some moral values are "universal," suggesting an absolutist position. But, in contrast to moral theories that derive human moral obligations ultimately from a "supreme being," Darwin's theories are derived from social utility. Codes of behavior that enlarge the possibilities for the survival of the human community at any stage of its development will be selected for survival.[10] Moral codes that, when followed, lead to social decay and dissolution will be weeded out. Volition, or free will, on this account of morality is ruled out. But the loss is not as great as one might imagine. Events have so conspired in the peculiar evolution of human societies to ensure that just those behaviors that would have been endorsed under a free will doctrine will still be endorsed under a social utility doctrine, and just those behaviors that would have been condemned under a free will doctrine will be condemned under a social utility doctrine. Darwin may have thrown out the bathwater of morality, but he did not throw out the baby.

THE GOD QUESTION

This excursion in Darwin's moral theorizing does not, however, settle the main issue that provoked the inquiry in the first place. That issue, it will be recalled, is whether

Darwin believed ultimately in a "random" universe, one that could dispense with the notion of a presiding divinity altogether. His move from a free will morality to a social utility morality does not answer this question. It may suggest that God is no longer necessary in the world, but it does not require that conclusion. It still may be the case that a divine Creator ordered human life, including natural human instincts, so as to ensure that, over the course of ages, humans would be protected against mutual self-destruction by the possession of the "right" instincts for survival. On this understanding, an intelligent Creator could be removed from the role of daily superintendence over human affairs, yet still be allowed the primeval role of putting the building blocks of life together in such a way as to ensure survival of humans even in a context of continuous geological and environmental change. Such a view, while drastically out of step with usual "Christian" conceptions of morality that continued to insist upon an irreducible human free volition as a *sine qua non* of genuine moral conduct, is nevertheless not unthinkable. Just as Darwin could have retained an original role for divine intelligence in the beginning to provide all living creatures with the "right" adaptive features in their natural organization to enable them to survive in a competitive world, so too might a role for creative intelligence have been preserved to equip humans in the beginning with the "right" social instincts to ensure their survival as well. Is this what Darwin thought?

Darwin was just not the sort of thinker who set out clear, straightforward answers to questions like this. Whenever he was uncertain, as he most evidently was in this case, he tended to ask himself questions rather than make assertions. No doubt when the young Darwin began his voyaging aboard the *Beagle* he took with him fairly conventional nineteenth-century British opinions about the deity. Yet it is equally clear that by the time he started his first transmutation notebook in 1837, *Notebook* B, in the face of unexpected discoveries in natural history, he had started to question old certainties:

> How does it come wandering birds, such as sandpipers, [are] not new at Galapagos? Did the creative force know that <<these>> species could arrive? Did it only create those kinds not so likely to wander? Did it create two species closely allied to Mus. Coronata, but not coronata? (B-100).

A similar question was raised further on in the B *Notebook*:

> Many trees Compositae, because seeds first arrived. <<Ferns ditto.>> Hence formed trees. & would creator <on volcanic island> *make* plants <grow closely>? When this volcanic point appeared in the great ocean, [would the creator] have made plants of American and African form, merely because [they occupy an] intermediate position? (B-193).

Darwin had no answer at this point. But he was already being drawn toward the idea of an absent deity, one who in the beginning had provisioned his creations with characteristics necessary for success and laws to ensure the proper fulfillment of his divine plan:

> Astronomers might formerly have said God ordered each planet to move in its peculiar destiny. In [the] same manner God orders each animal created with [a] certain form in [a] certain country, but how much more simple, & sublime power let attraction act according to certain laws [and] such [as above] are inevitable consequen[ce]. Let animal be created, then by the fixed laws of generation, such will be their successors (B-101).

To this point Darwin found plenty of room in his speculations for an intelligent creator. At the same time, he was starting to show increasing contempt for any account of Nature that relied upon direct divine intervention, even occasional, in nature's unfolding:

> Absolute knowledge that species die & others replace them—two hypotheses. Fresh creation is mere assumption, it explains nothing further. Points gained if any facts are connected [with this assumption] (B-104).

At B-111 Darwin was questioning Étienne Geoffroy Saint-Hilaire's assertion that the great variety of animals was "*created*," (Darwin's emphasis), suggesting in the same passage that G. St. Hilaire must have meant "propagated" rather than "created," a position Darwin could embrace.[11]

Similarly, in *Notebook* C:

> An Entomologist going into country & collecting thousand & tens of thousands New insects, perhaps scarcely one new family & no new orders,—Wonderful, partly explained on my theory. Otherwise mere fact creator chooses so to create (C-200).

In other words, asserting that "the creator made it so" begs the question. "[To say] distinct creation," Darwin urged, is "a mere statement, nothing is explained" (C-209e). Indeed, the whole notion of special creation for every variation seemed unnecessary and even somewhat ridiculous:

> The common Mushroom & other cryptogamic plants same in Australia & Europe.—If creation be absolute thing, the creation must take place only when

> creator [fore]sees the means of transport [will] fail.—Otherwise no relation between means of Transport & creation exists.—Pooh [in other words, pure nonsense]. May have been Created at many spots & since disseminated (C-240e).

Darwin, in short, was dissatisfied with an explanation that he regarded as a tautology: the world we experience is a consequence of a "creator who chooses to create" the world we experience. Darwin was starting to carve out what he regarded as a more exalted conception of the creator's role in producing life's pageantry than what he found in those naturalists who saw the divine hand in every slight alteration in nature's fabric. Such views as these latter, he claimed, were miserable and limited:

> Has the Creator since the Cambrian formations gone on creating animals with [the] same general structure [as they had in the beginning]? [This is a] miserable limited view [of the Creator] (B-216).

But what view of creation was Darwin to put in its place? He apparently understood he had two options: one, that no divine intelligence presides over the universe at all, a "new kind of atheism," as he called it in *Notebook* M (M-74). The other was that an intelligent creator does exist, and that he designed the universe to operate according to "fixed laws." The evidence on the whole from the *Notebooks* confirms much more the latter conclusion by Darwin than the former. As early as *Notebook* D Darwin already was referring to a Creator and his "laws," suggesting the impossibility that a creator of the laws governing the universe could violate the laws he had already made:

> <<[My grandfather's theory of Mules not [being] hereditary [because they lack the power] of generation>> [is] false. *The creator would thus contradict his own law* (D-19, emphasis in original).[12]

That laws, and not an ever-present intelligent presence, best account for the natural phenomena of change and speciation, now becomes a frequent theme in the *Notebooks* (e.g., B-43, C-70, C-122–3, C-166, D-49, D-65–6, D 69, D-100 n. 2, E-3, E-53, etc.). But whence these laws? It is pretty clear that Darwin's opinion through most of the notebooks was that "God" was the source of these laws. Such a view, he maintained, is far more in keeping with the supposed grandeur of God than a caricature that makes of God a petty meddler in everyday earthly affairs, constantly violating the very laws he had made to govern physical events:

> What a magnificent view one can take of the world. Astronomical <and unknown> causes modified by unknown ones cause changes in geography &

> changes of climate superadded to change of climate from physical causes.—These superinduce changes of form in the organic world, as adaptation, & these changing affect each other & their bodies. By certain laws of harmony [these forms] keep perfect in these themselves [and thus] the world peopled <<with myriads of distinct forms>> from a period short of eternity to the present time, to the future.—How far grander than idea from cramped imagination that God created [everything], (warring against those very laws he established in all organic nature... How beneath the dignity of him who <<is supposed to have>> said let there be light & there was light (D-36–7).

Evidence abounds, Darwin argued, that God works by rules, not by direct intervention:

> The extinction of the S. American quadrupeds is difficulty on any theory—without God is supposed to create & destroy without rule.—But what does he [do] in this world without [a] rule? The destruction of great Mammals over whole world shows there is rule (D-72; cf. D-74 and *CDN* 357, n. 74-1).

But can we be sure that Darwin believed God was the *source* of the laws that govern nature? His suggestion that God cannot violate his own laws, cited above, strongly suggests that he did. But all doubt on that score is removed in *Notebook* M:

> Those savages who thus argue [that thunder and lightning are the direct will of God] make the same mistake, more apparent however to us, as does that philosopher who says the innate knowledge of the creator <<has been>> implanted in us... by a separate act of God, & not as a necessary integral part of his most magnificent laws, of which we profane in thinking not capable to produce every effect of every kind which surrounds us (M-135–6).

Thunder and lightning, and by inference all natural phenomena, may be explained by reference to the operation of natural laws that God has fashioned. The same is true of human morality. In the beginning God deposited a "moral sense" in humans, then withdrew to allow humans to work out their own moral fate according to the intelligent configuration of this special instinct (M-151). Yet people are reluctant to allow such a distant role for the divinity:

> [The general] unwillingness to consider Creator as governing by laws is probably that as long as we consider each object an act of separate creation we admire it more, because we can compare it to the standard of our own minds, which

> ceases to be the case when we consider the formation of laws invoking laws, & giving rise at last even to the perception of a final cause (M-154).

In short, Darwin's "grand view" in the *Notebooks* was not that no original source of intelligent creation existed. Rather, a Creator did create, in the beginning, but the "magnificence" of his creation is that he established the universe on a plan of "laws of harmony" that would thenceforth ensure an orderly and reliable unfolding of a divine plan. Constant, daily interference by the Creator is, on this view, not only unnecessary, but also demeaning of his sublime grandeur (D-36).

DARWIN AFTER 1838

That would seem to be Darwin's final verdict on the question—until we get to the "Old and Useless Notes." Hints of atheistic leanings had come earlier, particularly in *Notebook* C, where Darwin had realized, perhaps for the first time, that the logic of his theory lead to the inevitable conclusion that humans are not special—just another twig on a branch of a limb on the tree of life. The revelation was not easy to come by and caused him evident pain. The passage is important and requires extensive quotation:

> The believing that monkey would breed (if mankind destroyed) some intellectual being though not MAN—is as difficult to understand as Lyell's doctrine of slow movements &c &c. This multiplication of little means & bringing the mind to grapple with great effect produced, is a most laborious, & painful effort of the mind (although this may appear an absurd saying) & will never be conquered by anyone (if has any kind of prejudices) who just takes up and lay down the subject without long meditation.... Once grant that <<species>> [of] one genus may pass into other... (if this be granted!!) & whole fabric totters and falls.... The fabric falls! But Man—wonderful Man. "divino ore versus coelum attentus" is an exception. He is mammalian, his origin has not been indefinite—he is not a deity, his end under present form will come (or how dreadfully we are deceived), then he is no exception... Present monkeys might not [make a man], but [with proper contingencies] probably would (C-74–8).

It is one thing, from the standpoint of a presumed divinely ordered universe, to deny that humans possess free will. It is also one thing to believe one can preserve divinity under a social utility doctrine, as we have seen. It is quite another to believe humans just happened to come along in a process of sprouting animal evolution because the "proper contingencies" happened to be in place to enable monkeys to develop larger brains. The latter scenario makes a divine presence at *any* stage of life's evolution much less necessary, if not dispensable altogether. But two other pieces of the fabric would have to fall to clinch the case.

The first is the idea of an immortal soul that could anticipate an afterlife of rewards and punishments. If there is a God, even an absent God, it would make sense to believe that he cared about humans and their behaviors, finding some behaviors to be more choice worthy than others; and that he rewarded preferred behaviors and punished bad ones. Such, in fact, were accepted conventions of theological thinking in Darwin's day, even among social utilitarians. Darwin appears to doubt such stories. If humans really do not have free will, the concept of punishment, either heavenly or earthly for that matter, as a "repayment" for sin makes no sense. Nor does the idea of reward as a payment for "goodness" make any sense either. As to earthly retribution or reward, Darwin allows a utilitarian defense:

> One must view a wicked man like a sickly one. It would however be more proper to pity than to hate. We ought to pity and assist and educate [the wicked] by putting contingencies in the way to aid motive power.—If incorrigibly bad nothing will cure him. Yet it is right to punish criminals, but solely to *deter* others ("OUN" 26–7 and "footnote opposite page," in *CDN* 608, emphasis in original).

But such reasoning does not work for divinely imposed penalties, or by extension, rewards. Telling *stories* about such matters may help induce the proper moral conduct, and in fact the utility of such stories in promoting social peace is a plausible explanation for why such stories were invented in the first place. But to believe such stories convey a real truth is to suppose an absurdity, that God would punish or reward people who are incapable of acting otherwise than they do.

> These views are directly opposed & inexplicable if we suppose that the sins of man are under his control, & that a future life is a reward or retribution.—It may be a consequence but nothing further ("OUN" 28).

The last sentence is ambiguous. It could mean that stories about an afterlife are a consequence of human wrongdoing, "but nothing further," that is, not to be taken literally. Or it could mean that it is *possible* there is an afterlife, but that it would not be one in which the prospects of punishment and reward are designed to bring sins "under human control." The latter is impossible *ex hypothesi*, so if reward or retribution awaits us after we die, it is only a "consequence" of our behavior, not a reward or retribution strictly speaking. The former reading seems more plausible.

THE ORIGIN OF LIFE

That reading is strengthened when we turn to consider the second part of the fabric, Darwin's views about the very origin of life itself. While Herschel may have thought the "origin of species" was the great "mystery of mysteries," in truth how life itself arose on this planet posed an even more profound puzzle, as Darwin no doubt understood. In the *Origin* he was at pains to insist that he was not after a solution to *that* problem (*Origin* VII.5 [*Variorum* 380]). But in the "OUN" he did raise the question, at least to himself, and came across a startling conclusion, and one that is perhaps even more prescient than his discovery of natural selection. Perhaps life itself, he suggested, was just the accidental coming together of bits of inorganic matter according to no fixed laws at all! This would indeed be an entryway into a Godless universe:

> Effects of Life in the abstract is matter united by certain laws different from those that govern the inorganic world; life itself being the *capability* of such matter obeying a certain & peculiar system of movements different from organic movements.... Has any vegetable or animal *matter* been formed by the union of *simple* non-organic matter, without action of vital laws—According to the individual forms of living beings, matter is united in different modification, peculiarities of external form impressed, and different laws of movement.... Organic laws probably have some unknown connection to [an inorganic principle].... In simplest forms of living being... vital laws act definitely (<<as>> chemical laws) as long as certain contingencies are present (as heat, light &c) ("OUN" 34).

The idea in this fascinating speculation seems to be that life itself arose as a purely "contingent" result of certain bits of inorganic matter, responding to local circumstances such as a certain amount of heat and light, uniting in a novel way by accident, not by law. The individual components of this inorganic matter may themselves have been under the influence of "laws of inorganic matter," but why or when or how the parts necessary for life came together was a pure matter of accident (a word Darwin employed three times in the sentences immediately following the quoted material).

Here, at last, we find Darwin's dangerous idea: the purely chance origin of life on this planet. Once chance governs the beginning, chance can govern all the way down. God has been ushered out the door, as being superfluous to any account of how life is as it is or how it has come to be as it has. The planet is on its own.

But Darwin did not go public with these views. As we have seen, he consigned them to a set of notes that he probably wished not to be seen by anyone else. He may have even succeeded in blocking these most radical implications of his theory of natural selection from his own mind. Even in these hidden notes, when he asks himself in effect, "do these views make me an atheist?," he responded with a vehement "NO!" ("OUN" 37 n. 37-3). Many years later, when he composed his *Autobiography*, he decided that the most fitting way to characterize his religious views was "theistic," then later "agnostic." Though not the same, both seem to preserve at least the possibility, if not the likelihood, of a Creator who designed a world in the beginning that would operate in definite and predictable ways. Whether Darwin took comfort in these self-appraisals is hard to know. We are at the borderland of a psychoanalytic analysis of this brilliant man, and that is a different study.

NOTES

1. The best recent biographies are unanimous in this verdict: Browne (1995, 2002); Desmond and Moore (1991, 2009). If Darwin has been faulted for any moral failing it is for taking key ideas of his theory from others without proper acknowledgment, but that view is no longer widely credited. Cf. Johnson (2007) for a review of the controversies.

2. This terrain has been surveyed carefully by R. J. Richards (1989, 110–42); Paul Farber (1994); and P. Barrett (in Gruber, 1981). These works must be the starting point for any serious study of the development of a Darwinian moral theory. Richards traces continuities from Darwin's *Notebooks* through his *Descent of Man*, finding, correctly, that Darwin did not really alter his views in that 35-year interval. Farber looks at how Darwin's thoughts on ethics influenced the subsequent development of a Darwinian ethics. I focus mainly on the *Notebooks*, the focus of Barrett's work. In keeping with the argument of the present work I find that what Darwin did change was his discussions about "chance" in human moral development. He did not change his opinion but again only his mode of exposition. "Chance" variation as an expression all but disappears in the *Descent* and the *Expression of Emotions*.

3. The "Old and Useless Notes" ("OUN") may be found in P. H. Barrett (1974), "Early Writings of Charles Darwin" along with Barrett's extensive annotations. Barrett's essay appears in H. E. Gruber (1981). A later transcription is Barrett et al. (1987).

4. Mackintosh's views and their influence on Darwin are discussed in R. J. Richards (1989, 116–20); and Barrett (in Gruber, 1981). Richards (n. 2 above) demonstrates that Darwin believed his theory could resolve an apparent conflict between the views of those who traced moral behavior in humans to an innate "impulse" (i.e., motive) for "being/doing good" and those who believed it rested upon an innate faculty of "moral judgment." Why, Darwin wondered, does the moral judgment seem invariably to affirm what the motive for morality commands? The faculty of judgment "knows," whereas the impulse for behavior "commands," yet the two coincide. Richards also argues persuasively that Darwin resolved this puzzle by establishing an evolutionary connection between impulse and judgment: moral behavior initially comes from the pleasure moral agents receive, first from having good being done to them, and then later (by association of ideas) doing good to others. Over generations the "habits" of goodness thus inculcated become "instinctual" by the Lamarckian mechanism of use-inheritance as applied to human moral conduct. I argue here that Darwin recognized earlier than Richards allows that "natural selection" of "chance variations" accounted for the human acquisitions of morality.

Richards (208, 238–9) summarizes what he takes to be Darwin's mature view of the evolution of morality.

5. Richards argues that the "pleasure drive" as the motive for moral behavior was replaced (by natural selection) by an acquired instinct that promoted morally approvable conduct, enforced by a mechanism that he calls "community selection" (1989, 212–9), a variant of the more familiar "group selection" idea of other evolutionary theorists. (For Darwin, "group selection" was limited to families and tribes—sometimes called "kin selection," not "society in general.") I grant that Darwin identified "group selection" as an important mechanism in evolutionary change, but the individual person acting from motives that promote the good of the group also receives pleasure for himself, and the anticipation of that pleasure is no doubt what drives the conduct. I do not dispute that such conduct also has beneficent consequences to the community, so I doubt my disagreement with Richards on this point runs very deep.

6. This point was made by R.J. Richards (1989, 116–123).

7. See previous two notes for R. J. Richards' clarifications.

8. Martineau's views and Darwin's response to them are discussed in R. J. Richards (1989, 112–14).

9. The editors of the *CDN* note that Darwin referred to the practice of cannibalism among the natives of Tierra del Fuego in *Origin*, 36, but then claim "this belief was later denied" (*CDN* 555, n. 142-2). But it was not denied here, or later, by Darwin, as the entry implies. It was instead denied by another voyager, a missionary in fact, E. L. Bridges, who had lived among the Fuegians in 1871. He claimed that Darwin "misunderstood his informants," in other words, that Darwin had not seen cannibalism firsthand. Thus, Darwin's belief that cannibalism existed in Tierra del Fuego remained part of his belief system probably to the end of his life.

10. In terms of a supposed controversy about whether Darwin was an "organism" selectionist or a "group" selectionist (cf. S. J. Gould, 2002), Darwin's account of human morality makes him a decidedly "group" selectionist. Society, not the individual, benefits from the "moral sense." But if one asks what motivates the actions that benefit the community, it seems clearly to be the moral agent's desire for personal pleasure. Darwin's commitment to group selectionism in at least some cases is well illustrated in the *Origin* and elsewhere by his treatment of "social insects" like bees and ants. Cf. R. J. Richards (1989, 212–9). Human behavior that arises solely from the desire to benefit others is the subject of an extended controversy in contemporary biological theory that may be summarized under the question, "Can humans be altruists?" An examination of the recent literature on this subject would take us too far afield, but cf. E. Sober (2000) for an introduction.

11. The passage from G. St. Hilaire that Darwin was referring to here is quoted at length in *CDN* 197, n. 111-1. But cf. Darwin's marginal notations to É. Geoffroy St. Hilaire (1830) and to I. Geoffroy St. Hilaire (1847). It is not clear when he read these works, but it was presumably after he composed the B *Notebook* because now Darwin had come to understand that Étienne was *not* a transmutationist (*Marginalia* 302, 320).

12. Erasmus Darwin's views about the infertility of mules are quoted in *CDN* 171, n. 2-2. The larger question about Darwin's views concerning the role of "secondary causes" (i.e., laws) that govern nature's working has been thoroughly discussed in the secondary literature, including the question whether Darwin regarded these secondary causes as "designed" by an omniscient God. I have earlier argued (chapter 3 of the present work) that Darwin did not really believe after 1838 that God designed the secondary causes. They, too, are "by chance." Discussions include Moore (1979, 256, 265, 274, 321, and 482); Hull (1973, 53, 60); Ruse (1979, 67–9, 83, 87–93).

Appendix

THE PRIMARY SOURCES

The secondary literature on Darwin is vast, and I leave it to the citations and notes to show my inestimable debts to other scholars. But even Darwin's own writings cover a great deal of territory. Any discussion of Darwin's thought, especially as concerns its evolution over the course of a long life, must take cognizance of the fact that the expression of his ideas is complexly layered. It is not enough to say that he had a "private" and a "public" side (Herbert, 1977, 179–80; Rudwick, 1982, 186–206; Browne, 1982, 275–80). His public writings—those that he wrote for publication—were often in themselves not stable. We have already mentioned the six editions of the *Origin* and the changes made by Darwin to each successive edition after the first. The same is true of several other works he published; most of them underwent multiple editions, always accompanied by revisions in wording from one to the next. Worthy of mention in this regard are the *Journal of Researches* (first edition 1839, second edition 1845); the book on orchids (first edition 1862, second edition 1877); and the *Variation of Animals and Plants Under Domestication* (first edition 1868, second edition 1875). Even his more private reflections, from the *Notebooks* through the 1844 "Essay," show frequent revision, no doubt in anticipation that eventually his key ideas would be published in the future (Kohn and Stauffer, 1982). As we said at the outset, "Darwinism," whatever it is, is a moving target, even with respect to published writings.

A second "public" layer includes books and papers Darwin published or intended to publish but that he did not alter after the first composition. Many of his shorter works, written over many years, have been published together in the *Collected Papers of Charles Darwin* (Paul Barrett, ed., 1977). R. Stauffer, ed., published Darwin's *Natural Selection* in 1975. This was the large manuscript that Darwin had initially intended as the definitive treatment of his subject on the origin of species, but that he decided to set aside in 1858 (after receiving Wallace's manuscript) in favor of producing a shorter "abstract," the work that in 1859 was published as the *Origin of Species*.

He at times had thoughts of completing the "big species book" after the *Origin* appeared, but settled instead upon producing other works that took up separate parts of the work that is now known as *Natural Selection*. Darwin also published several additional books on narrower topics, a full bibliography of which is available in Freeman (1977).

Complicating matters is Darwin's "private" side. Often this is understood to include his private correspondence (now in 19 volumes and counting through the Cambridge University "Darwin Correspondence Project") as well as other notes, manuscripts, and marginalia that he accumulated over many years. In fact these represent different layers of his private side. While no doubt he did not expect his letters or his other private notes ever to be published, these several sources must be treated differently, in terms of how to evaluate them. Letters, by definition, are written for an external audience, even if of one person only (and Darwin could feel assured that his letters would seldom be read by anyone but the person to whom they were addressed). But some of these letters must be regarded as more "private" than others. He would, for example, be more candid about his views with his closest collaborator J. D. Hooker than with, say, correspondents from Germany or Italy whom he had never met in person. Interlocutors such as Charles Lyell and Asa Gray would fall in-between these two extremes—much closer to Hooker among Darwin's correspondents than people he had never met, but still not quite on the same level of confidentiality as Hooker. These judgments are, of course, to some extent inferences: Darwin does not provide a roadmap. But everyone knows that some intimates are more intimate than others, and having some well-grounded thoughts about the pecking order alerts us to some sensitivities about how to read and interpret the letters.

Finally we come to the *Notebooks*, private notes, and marginalia. This group of writings constitutes an entirely new order of "privacy." Darwin composed them for no one other than himself: these were "for his eyes only," or, as we might say today, "confidential." Most of the material from the *Notebooks* dates from 1836–1842, and includes the several transmutation notebooks (Kohn et al., 1984) and the hastily written "Sketch" of the species theory of 1842. (The latter formed the basis of the submission to the Linnean Society Meeting of July 1, 1858, of Darwin's contribution, alongside Wallace's, to the species question. In a footnote Darwin noted, "This MS. work was never intended for publication, and therefore was not written with care" [Darwin, 1971 (1858), 257 n. 1]).

A second vast resource of "private" Darwin materials is the many notes and short manuscripts that exist in various file folders in several libraries around the world. Many of these are mere jottings—scraps of paper written down and tucked away by Darwin for future reference. This material is now being brought to public light by the efforts of David Kohn, director of the Darwin Manuscripts Project, and his associates at the American Museum of Natural History. According to the website home page, "the AMNH Darwin Manuscripts Project is a historical and textual edition of Charles Darwin's scientific manuscripts, designed from its inception as an online project. The database at its core—DARBASE—catalogues some 96,000 pages of Darwin scientific manuscripts. These are currently represented by 16,094 high resolution digital images. Thus far 9,871 manuscript pages have been transcribed to exacting standards and all are presented in easy to read format."

The marginalia are somewhat different. This body of writings is to be found in the margins of books and articles Darwin read over a long period of time, comprising everything from the first books he read to the last—a unique chronicle of the reading life of a single person. The marginalia are thus useful not just for seeing Darwin's most private thoughts but also, and in part because the annotations themselves are frequently "layered"—that is, composed at different times even for individual works—for tracing the evolution of his thought. Moreover, these writings constitute an immense library of material just by themselves. Only a small fraction of this material is relevant to the present work, but that fraction too is indispensable for getting at the question of what "Darwinism" may have meant to Darwin. The marginalia appear in a volume composed and edited by Di Gregorio and Gill (1990).

All of this material presents hermeneutical challenges (cf. Hodge and Kohn, in Kohn, 1985, 185). One question is, how important is this work? The *Notebooks* and "Sketch" were written when Darwin was a young man whose ideas for the most part were not yet fully developed. That could cause a critic to dismiss their value for understanding the mature Darwinian theory. The marginalia consist of abbreviated jottings, the underlining of text, and the scoring of passages in reference to works written by others, and so are sometimes difficult to decipher, and the same is true of much in the DARBASE. But most scholars believe that to dismiss this work as unimportant would be a mistake. If nothing else, it gives us a picture of how Darwin's thought apparently evolved. That by itself is an important consideration for viewing the early, most private writings with seriousness.

By good fortune, almost all of Darwin's "private writings," including his correspondence down to 1871 and a number of letters written after that, have now been published, mainly in the *Correspondence of Charles Darwin* (1985–2012) (see also the several volumes of *Life and Letters of Charles Darwin*, published much earlier by Darwin's son Francis). The *Notebooks* are available in several different editions, a useful one of which is *Charles Darwin's Notebooks* (1985), edited by Kohn and Barrett. The 1842 "Sketch" and the 1844 "Essay" are available in a 1971 volume published by Cambridge University Press as *Evolution By Natural Selection*, edited by Gavin de Beer. This volume reprints earlier copies of both works that were originally produced for publication in 1909 by Darwin's son Francis. This volume also includes the papers of Darwin and Wallace that were published in the *Journal of the Linnean Society of London* in 1858. Charles Darwin's *Autobiography*, which originally appeared in a "sanitized" version edited by his family, is now available in an unabridged form, which was produced for publication by Darwin's granddaughter Nora Barlow (1958).

Finally, an indispensable resource for Darwin studies is the *Variorum* edition of the *Origin*, published by Morse Peckham in 1959. Darwin's masterpiece appeared in six editions, each different from the others. Editions 2–5 are rare and have seldom been reprinted after their first issues. The two editions most people read are the first and/or the sixth, both of which have been frequently reprinted. The Peckham volume brings all six editions together within the compass of a single volume.

If one accepts the premise of the current study that Darwin regarded his "breakthrough" in evolutionary thinking to center on the idea of "chance" in evolution, these early private writings are of the utmost significance. They should show when, and why, Darwin came to embrace "chance" in his theory. And they do. Some of this spadework has already been undertaken by other scholars, especially in several essays in the edited volume *Darwinian Heritage* (D. Kohn, 1985, especially Hodge and Kohn, "Immediate Origins of Natural Selection"), and in several recent studies by John Beatty and James Lennox.

This work now needs to be supplemented by additional discoveries, some of which I have presented in the main text of the present work. The *Notebooks*, for example, show that Darwin came across chance earlier than Hodge and Kohn recognize, not so much in terms of how variations come about by chance in Darwin's view, but how he may have been led to this important idea in the first place. Evidently, Darwin first stumbled upon "chance" in nature as a direct result of reading Charles Lyell's *Principles of Geology* in 1831–1832. The context is not so much "chance variations" as it is "chance transport" (see chapter 2). How do organisms get from one favorable environmental situation to another? Lyell said, several times, "by chance," and Darwin agreed. But "chance transport" is almost unthinkable outside the context of "chance features" that enable such transport to occur. It may be—although this is not certain—that Darwin's own thoughts were "transported" from the realization that "transport" is often accidental, to the further thought that structural features that enable transport to occur are also accidental. Thus Lyell, always standing against chance variation in natural productions, may ironically have been Darwin's source for his famous idea.

BIBLIOGRAPHY

Agassiz, L. 1859. *An Essay on Classification*. 4 vols. London: Longman, Brown, Green, Longmans, & Roberts. Original edition, 1857–1862. Reprint, Contributions to the Natural History of the United States of America.

Al-Zahrani, A. 2008. "Darwin's Metaphors Revisited: Conceptual Metaphors, Conceptual Blends, and Idealized Cognitive Models in the Theory of Evolution." *Metaphor and Symbol* no. 23 (1):50–82.

Arnhart, L. 1984. "Darwin, Aristotle, and the Biology of Human Rights." *Social Science Information* no. 23:493–521.

Ashworth, J. H. 1935. "Charles Darwin as a Student in Edinburgh, 1825–1827." *Proceedings of the Royal Society of Edinburgh* no. 55:97–113.

Austin, J. L. *How to do Things with Words: The William James Lectures delivered at Harvard University in* 1955, 1962 (ed. J. O. Urmson and Marina Sbisà), Oxford:Clarendon Press.

Barrett, P. H. 1973. "Darwin's Gigantic Blunder." *Journal of Geological Education* no. 21 (1):19–28.

——. 1980. *Metaphysics, Materialism, and the Evolution of Mind: Early Writings of Charles Darwin*. Chicago: University of Chicago Press.

Barrett, P. H., Sandra Herbert, David Kohn, and Sidney Smith, eds. 1987. "Old and Useless Notes." In *Charles Darwin's Notebooks 1836–1844*, 599–629. Ithaca, NY, and London: British Museum and Cornell University Press.

Barthélemy-Madaule, M. 1982. *Lamarck: The Mythical Precursor*. Translated by M. H. Shank. Cambridge, MA: MIT Press.

Beatty, J. 1984. "Chance and Natural Selection." *Philosophy of Science* no. 51 (2):183–211.

——. 1987. "Dobzhansky and Drift: Facts, Values, and Chance in Evolutionary Biology." In *The Probabilistic Revolution*, edited by Kreuger et al. Cambridge, MA: MIT Press, vol. 2, Chapter 11.

——. 2006. "Chance Variation: Darwin on Orchids." *Philosophy of Science* no. 73 (5): 629–41.

——. 2010. Edited by Mueller and Pigliucci. Cambridge: MIT Press.

Beddal, B. G. 1988. "Darwin and Divergence: The Wallace Connection." *Journal of the History of Biology* no. 21 (1):1–68.

——. forthcoming. "The Face of Nature: Anthropomorphic Elements in Darwin's Style." In *The Language of Nature*, edited by L. Jordanova.

Bowler, P. J. 1983. *The Eclipse of Darwinism: Anti-Darwinian Evolution Theories in the Decades around 1900*. Baltimore: Johns Hopkins University Press.

Brandon, R. 1990. *Adaptation and Environment*. Princeton, NJ: Princeton University Press.

——. 1992. *The Eclipse of Darwinism: Anti-Darwinian Evolution Theories in the Decades around 1900*. Baltimore: Johns Hopkins University Press.

——. 1993. *Darwinism, Issue 6 of Twayne's Studies in Intellectual and Cultural History*. New York: Twayne Publishers.

Brent, P. L. 1981. *Charles Darwin: "A Man of Enlarged Curiosity."* London: Heinemann.

Brooke, J. H. 1991. *Science and Religion: Some Historical Perspectives*. Cambridge: Cambridge University Press.

Brooks, J. L. 1984. *Just before the Origin: Alfred Wallace's Theory of Evolution*. New York: Columbia University Press.

Browne, J. 1980. "Darwin's Botanical Arithmetic and the Principle of Divergence, 1854–1858." *Journal of the History of Biology* no. 13 (1):53–89.

——. 1982. "Essay Review: New Developments in Darwin Studies?" *Journal of the History of Biology* no. 15:275–80.

——. 2006. *Darwin's Origin of Species: A Biography*. New York: Atlantic Monthly Press.

——. 1985. "Darwin and the Face of Madness." In *The Anatomy of Madness: Essays in the History of Psychiatry*, edited by W. F. Bynum, R. Porter, and M. Shepherd, 151–65. London: Tavistock Publications.

——. 1995. *Charles Darwin: Voyaging*. Princeton, NJ: Princeton University Press.

——. 2002. *Charles Darwin: Power of Place*. Vol. 2. New York: Knopf.

Burkhardt, F., S. Smith, D. Kohn, W. Montgomery, C. Darwin, and American Council of Learned Societies. 1985. *A Calendar of the Correspondence of Charles Darwin, 1821–1882, Garland Reference Library of the Humanities*. New York: Garland Pub.

Burkhardt, R. W. 1977. *The Spirit of System: Lamarck and Evolutionary Biology*. Cambridge, MA: Harvard University Press.

Campbell, J. A. 1989. "The Invisible Rhetorician: Charles Darwin's "Third Party" Strategy." *Rhetorica: A Journal of the History of Rhetoric* no. 7 (1):55–85.

Carpenter, W. B. 1851. *Principles of Physiology, General and Comparative*. Philadelphia: Lea and Blanchard.

——. 1856. *Researches on the Foraminifera*. Vol. 146, *Philosophical Transactions of the Royal Society of London*. London: The Royal Society.

Corsi, P. 1988. *The Age of Lamarck: Evolutionary Theories in France, 1790–1830*. Rev. and updated ed. Berkeley: University of California Press.

——. 1988. *Science and Religion: Baden Powell and the Anglican Debate, 1800–1860*. Cambridge: Cambridge University Press.

Cuvier, F. 1827. "Essay on the Domestication of Mammiferous Animals." *Edinburgh New Philosophy Journal* no. 3, 303–18; no. 4, 45–60; 292–8.

Cuvier, F., G. Cuvier, and H. M'murtrie. 1833. *The Animal Kingdom: Arranged in Conformity with Its Organization; Translated from the French, and Abridged for the Use of Schools by H. M'murtrie*. New York: G. & C. & H. Carvill.

Cuvier, G., and P. A. Latreille. 1834. *The Animal Kingdom, Arranged According to Its Organization, Serving as a Foundation for the Natural History of Animals, and an Introduction to Comparative Anatomy*. 4 vols. London: G. Henderson.

Dagg, A. I., and J. B. Foster. 1976. *The Giraffe: Its Biology, Behavior, and Ecology*. New York: Van Nostrand Reinhold Co.

Darwin, C. 1859. *On the Origin of Species by Means of Natural Selection, or the Preservation of Favored Races in the Struggle for Life*. London: Murray (see also Peckham, M, 1959).

——. 1885. *The Variation of Animals and Plants under Domestication*. 2nd ed. London: Murray.

——. 1887. *The Life and Letters of Charles Darwin, Including an Autobiographical Chapter*, edited by F. Darwin. 2 vols. London: John Murray.

——. 1958. *Autobiography. With Original Omissions Restored*. London: Collins.

——. 1964. *On the Origin of Species: A Facsimile of the First Edition*. Cambridge: Harvard University Press. (See also Peckham, M., 1959)

——. 1981. *The Descent of Man, and Selection in Relation to Sex*. Princeton, NJ: Princeton University Press.

——. 1985–2012. *The Correspondence of Charles Darwin*, edited by S. S. Frederick Burkhardt. 19 vols. Cambridge: Cambridge University Press.

——. 1987. *Charles Darwin's Notebooks 1836–1844*. Edited by P. Barrett, Sandra Herbert, David Kohn, and Sidney Smith. Ithaca, NY, and London: Cornell University Press.

——. 1997. "Origin of the Species." In *Lightbinders, Inc. CD-ROM edition of Darwin's works*, edited by P. Goldie and M. Ghiselin. San Francisco: Lightbinders, Inc.

Darwin, C., E. J. Browne, and M. Neve. 1989. *Voyage of the Beagle: Charles Darwin's Journal of Researches*. Penguin Classics. London: Penguin Books.

Darwin, C., and J. W. Burrow. 1985. *The Origin of Species by Means of Natural Selection, or, the Preservation of Favored Races in the Struggle for Life*. London: Penguin Books.

Darwin, C., and J. T. Costa. 2009. *The Annotated Origin: A Facsimile of the First Edition of on the Origin of Species*. Cambridge, MA: Belknap Press of Harvard University Press.

Darwin, C., M. A. Di Gregorio, and N. W. Gill. 1990. *Charles Darwin's Marginalia, Garland Reference Library of the Humanities*. New York: Garland.

Darwin, C., J. S. Henslow, and N. Barlow. 1967. *Darwin and Henslow: The Growth of an Idea; Letters, 1831–1860*. London: Murray for Bentham-Moxon Trust.

Darwin, C., and M. Peckham. 1959. *The Origin of Species by Charles Darwin: A Variorum Text*. Philadelphia: University of Pennsylvania Press.

Darwin, C., and R. C. Stauffer. 1975. *Charles Darwin's Natural Selection: Being the Second Part of His Big Species Book Written from 1856 to 1858*. Cambridge: Cambridge University Press.

Darwin, C., and A. R. Wallace. 1958. *Evolution by Natural Selection*. Cambridge: Pub. for the XV International Congress of Zoology and the Linnean Society of London at the University Press.

——. 1971. *Evolution by Natural Selection*. Gavin de Beer, ed. Cambridge: Cambridge University Press.

Darwin, E. 1801. *Zoonomia; or, the Laws of Organic Life*. 3d ed. 4 vols. London: J. Johnson.

Darwin, Francis, ed. 1959. *Life and Letters of Charles Darwin*, 2 vols. New York: Basic Books.

Dawkins, R. 1986. *The Blind Watchmaker*. Harlow: Longman.

Dawkins, Richard. 1989. *The Selfish Gene*, 2nd ed. New York: Oxford University Press.

——. 2003. *A Devil's Chaplain*. Boston and New York: Houghton and Mifflin.

——. 2006. *The God Delusion*. Boston and New York: Houghton Mifflin Co.

De Beer, Gavin, ed. 1971. *Evolution by Natural Selection*. Cambridge: Cambridge University Press.

Dempster, W. J. 1983. *Patrick Matthew and Natural Selection: Nineteenth Century Gentleman-Farmer, Naturalist, and Writer*. Edinburgh: P. Harris.

——. 1996. *Evolutionary Concepts in the Nineteenth Century: Natural Selection and Patrick Matthew*. Durham, NC: Pentland Press.

——. 2005. *The Illustrious Hunter and the Darwins*. Sussex: Guild Publishing.

Dennett, Daniel. 1995. *Darwin's Dangerous Idea: Evolution and the Meanings of Life*. New York: Simon and Schuster.

——. 2006. *Breaking the Spell: Religion as a Natural Phenomenon*. New York: Viking.

Desmond, A., and J. Moore. 1991. *Darwin*. London: Michael Joseph.

Desmond, A. and J. Moore. 2009. Darwin's Sacred Cause: How a Hatred of Slavery Shaped Darwin's Views on Human Evolution. Boston & New York: Houghton, Miflin, Harcourt.

Desmond, A. J. 1989. *The Politics of Evolution: Morphology, Medicine, and Reform in Radical London, Science and Its Conceptual Foundations*. Chicago: University of Chicago Press.

Dilley, Steven. 2013. *Darwinian Evolution and Classical Liberalism*. Lanham, Maryland: Lexington.

Dobzhansky, T. 1959. "Blyth, Darwin, and Natural Selection." *The American Naturalist* no. 93 (870):204–06.

Dyson, Freeman. 2001. "Review of Recent Literature on Intelligent Design." *New York Review of Books*.

Eble, G. 1999. "On the Dual Nature of Chance in Evolutionary Biology and Paleobiology." *Paleobiology* no. 25:75–87.

Egerton, F. N. 1976. "Darwin's Early Reading of Lamarck." *Isis* no. 67:452–6.

Eiseley, L. 1956. *Darwin's Century: Evolution and the Men Who Discovered It*. Garden City, NY: Doubleday.

——. 1959. "Charles Darwin, Edward Blyth, and the Theory of Natural Selection." *Proceedings of the American Philosophical Society* no. 103 (1):94–158.

——. 1972. "The Intellectual Antecedents of the Descent of Man." In *Sexual Selection and the Descent of Man*, edited by B. Campbell, 1–16. Chicago: Aldine.

——. 1959. *Darwin's Century: Evolution and the Men Who Discovered It*. London: Gollancz.

——. 1965. "Darwin, Coleridge, and the Theory of Unconscious Creation." *Daedalus* no. 94 (3):588–602.

Evans, L. T. 1984. "Darwin's Use of the Analogy between Artificial and Natural Selection." *Journal of the History of Biology* no. 17 (1):113–40.

Farber, Paul. 1994. *Temptations of an Evolutional Ethics*. Berkeley: University of California Press.

Fleming, J. 1882. *Philosophy of Zoology; or, a General View of the Structure, Functions, and Classification of Animals*. 8 vols. Edinburgh.

Fontes da Costa, Palmira. 2010. "Darwin's Notebooks on the Moral Sense of Man." *Antropologia Portuguesa* 27, 137–47.

Fox Keller, E., and E. Lloyd, eds. 1992. *Keywords in Evolutionary Biology*, Cambridge MA: Harvard University Press.

Freeman, R. B. 1977. *The Works of Charles Darwin: An Annotated Bibliographical Handlist*. 2nd ed. Dawson: Folkstone.

Galton, F. 1874. *English Men of Science: Their Nature and Nurture*. London: Macmillan & Co.

——. 2003, "From Darwin to Today in Evolutionary Biology," in Hodge and Radick, 240–64.

Gayon, J. 2005. "Chance, Explanation, and Causation in Evolutionary Theory." *History and Philosophy of the Life Sciences* 27:395–405.

Ghiselin, M. 1969. *The Triumph of the Darwinian Method*. California: University of California Press.

Ghiselin, M. *Metaphysics and the Origin of Species*. State Uiversity of New York Press.

Gigerenzer, G. 1989. *The Empire of Chance: How Probability Changed Science and Everyday Life, Ideas in Context*. Cambridge: Cambridge University Press.

Gildenhuys, Peter. 2004. "Darwin, Herschel, and the Role of Analogy in Darwin's *Origin*." *Studies in History and Philosophy of Science* no. 35:593–611.

Gillespie, N. 1979. *Charles Darwin and the Problem of Creation*. Chicago: University of Chicago Press.

Gillispie, C. C. 1968. "Lamarck and Darwin in the History of Science." In *Forerunners of Darwin, 1745–1859*, edited by B. Glass, 265–91. Baltimore: John Hopkins University Press.

Glass, B. 1959. "Maupertius, Pioneer of Genetics and Evolution." In *Forerunners of Darwin: 1745–1859*, edited by B. Glass, 51–83. Baltimore: John Hopkins University Press.

Glass, B., O. Temkin, and J. William L. Straus. 1959. *Forerunners of Darwin, 1745–1859*, edited by B. Glass et al., Baltimore: Johns Hopkins University Press.

Glick, T. F. 1974. *The Comparative Reception of Darwinism*. Austin: University of Texas Press.

Gotthelf, A. 1999. "Darwin on Aristotle." *Journal of the History of Biology* no. 32 (1):3–30.

Gould, S. J. 1982. "Darwinism and the Expansion of Evolutionary Theory." *Science* no. 216:380–7.

Gould, S. J. 1996. "The Tallest Tale: Is the Textbook Version of Giraffe Evolution a Bit of a Stretch?" *Journal of Natural History* no. 105 (5):18–26.

——. 2002. *The Structure of Evolutionary Theory*. Cambridge, MA: Belknap Press of Harvard University Press.

Gray, A. 1868. Review of Charles Darwin, *Variation of Animals and Plants under Domestication*. *The Nation*, 234–6.

——. 1876. *Darwiniana: Essays and Reviews Pertaining to Darwinism*. New York: D. Appleton and Company.

Greene, J. C. 1992. "From Aristotle to Darwin: Reflections on Ernst Mayr's Interpretation in 'The Growth of Biological Thought.'" *Journal of the History of Biology* no. 25 (2):257–84.

Grene, M., and D. Depew. 2004. *The Philosophy of Biology: An Episodic History*. Cambridge: Cambridge University Press.

Gruber, H. E. 1977. "The Fortunes of a Basic Darwinian Idea: Chance." *Annals of the New York Academy of Sciences* no. 291 (1):233–45.

——. 1981. *Darwin on Man: A Psychological Study of Scientific Creativity*. 2nd ed. Chicago: Chicago University Press.

Gruber, H. E., C. Darwin, and P. H. Barrett. 1974. *Darwin on Man: A Psychological Study of Scientific Creativity*. 1st ed. New York: E. P. Dutton.

Gruber, J. W. 1969. "Who Was the Beagle's Naturalist?" *The British Journal for the History of Science* no. 4 (3):266–82.

Guilding, L. 1825. "Description of a New Species of Onchidium." *Transactions of the Linnean Society of London* no. 14:322–4.

Henslow, J. S. 1833. *Syllabus of a Course of Lectures on Botany*. Cambridge: Privately Printed.

——. 1836. "Descriptive and Physiological Botany." In *The Cabinet Cyclopedia*, edited by D. Lardner. London: Longman, Rees, Orme, Brown, Green and Longman.

——. 1837. "Description of Two New Species of Opuntia: With Remarks on the Structure of the Fruit of Rhipsalis." *Magazine of Zoology and Botany*: 466.

Herschel, J. F. W. 1831. *A Preliminary Discourse on the Study of Natural Philosophy, The Cabinet Cyclopaedia Natural Philosophy*. London: Printed for Longman, Rees, Orme, Brown, Green and John Taylor.

Herbert, S. 1971. "Darwin, Malthus, and Selection." *Journal of the History of Biology* no. 4 (1): 209–17.

——. 1974. "The Place of Man in the Development of Darwin's Theory of Transmutation, Part 1." *Journal of the History of Biology* no. 7 (2):217–58.

——. 1977. "The Place of Man in the Development of Darwin's Theory of Transmutation, Part 2." *Journal of the History of Biology* no. 10 (2):155–227.

——, ed. 1980. *The Red Notebooks of Charles Darwin*. Ithaca, NY, and London: British Museum and Cornell University Press.

——. 1982. "Remembering Charles Darwin as a Geologist." In *Charles Darwin 1809–1882: A Centennial Commemorative*, edited by R. G. Chapman. Wellington, New Zealand: Nova Pacifica.

Hesse, M. B. 1966. *Models and Analogies in Science*. Notre Dame, IN: University of Notre Dame Press.

Hitching, F. 1982. *The Neck of the Giraffe, or, Where Darwin Went Wrong*. New Haven, CT: Ticknor and Fields.

Hobbes, Thomas. 1651. *Leviathan*. London.

Hodge, M. J. S. 1977. "The Structure and Strategy of Darwin's 'Long Argument.'" *British Journal of the History of Science* no. 10 (3):237–45.

——. 1982. "Darwin and the Laws of the Animate Part of the Terrestrial System (1835–1837): On the Lyellian Origins of His Zoonomical Explanatory Program." *Studies in History of Biology* no. 7:1–106.

——. 1983. "The Development of Darwin's General Biological Theorizing." In *Evolution from Molecules to Men*, edited by D. S. Bendall, 43–62. Cambridge: Cambridge University Press.

——. 1987. "Natural Selection as a Causal, Empirical, and Probabilistic Theory." In *The Probabilistic Revolution*, edited by Kreuger et al., vol. 2, chapter 10. Cambridge, MA: MIT Press.

——. 2003. Hodge, J., and G. Radick, eds. *The Cambridge Companion to Darwin*, Cambridge: Cambridge University Press.

Hog, E. 2004. "The Depth of the Heavens: Belief and Knowledge during 2500 Years." *Europhysics News*.

Hull, D. L. 1973a. *Darwin and His Critics; the Reception of Darwin's Theory of Evolution by the Scientific Community*. Cambridge, MA: Harvard University Press.

——. 1973b. "Charles Darwin and Nineteenth Century Philosophies of Science." In *Foundations of Scientific Method: The Nineteenth Century*, edited by R. N. Giere and R. S. Westfall, 115–32. Bloomington: Indiana University Press.

Humboldt, A. V., and A. Bonpland. 1819–1829. *Personal Narrative of Travels to the Equinoctial Regions of America, During the Years 1799–1804*. 7 vols. London: Longman, Hurst, Rees, Orme and Brown.

Hume, D. 1779. *Dialogues Concerning Natural Religion*. 2nd ed. Oxford: Oxford University Press.

Johnson, C. 2007. "The Preface to Darwin's Origin of Species: The Curious History of the Historical Sketch." *Journal of the History of Biology* no. 40 (3):529–56.

——. 2010. *The Idea of 'Chance' in Darwin's Thought., Proceedings of a Symposium of 150 Years of Darwin's Impact on the Social Sciences and Humanities*. San Diego: SDSU Press.

Jones, G. 1978. "The Social History of Darwin's Descent of Man." *Economy and Society* no. 7 (1):1–23.

Jordanova, L. J. 1984. *Lamarck, Past Masters*. Oxford: Oxford University Press.

Judd, J. W. 1911. "Charles Darwin's Earliest Doubts Concerning the Immutability of Species." *Nature* no. 1292 (88):8–12.

Kass, L. R. 1978. "Teleology and Darwin's the Origin of Species: Beyond Chance and Necessity?" In *Organism, Medicine, and Metaphysics: Essays in Honor of Hans Jonas on His 75th Birthday*, edited by S. F. Spicker and H. Jonas, 97–120. Dordrecht: D. Reidel Pub. Co.

Keyt, D. 1987. "Three Fundamental Theorems in Aristotle's *Politics*." *Phronesis* 32, 54–79.

——. 2011. "Aristotle's Political Theory." *Stanford Encyclopedia of Philosophy*.

Kimura, M. 1981. "The Neutral Theory as a Basis for Understanding the Mechanism of Evolution and Variation at the Molecular Level." In *Molecular Evolution, Protein Polymorphism and the Neutral Theory*, edited by M. Kimura, 363. New York: Springer.

——. 1992, "Neutralism," in *Keywords in Evolutionary Biology*, edited by Fox Keller and Lloyd, 225–30. Cambridge MA: Harvard University Press.

Kirby, W., W. Spence, and E. O. Essig. 1826. *An Introduction to Entomology: Or Elements of the Natural History of Insects: With Plates*. London: Longman.

Kohn, D. 1980. "Theories to Work by: Rejected Theories, Reproduction, and Darwin's Path to Natural Selection." *Studies in History of Biology* no. 4:67–170.

——. 1986. "The Darwinian Heritage." *History of the Behavioral Sciences* no. 24 (4): 413–15.

Kohn, D., R. C. Stauffer, and S. Smith. 1982. "New Light on "the Foundations of the Origin of Species": A Reconstruction of the Archival Record." *Journal of the History of Biology* no. 15 (3):419–42.

Krueger, L., G. Gigerenzer, and M. Morgan, eds. 1987. *The Probabilistic Revolution*. 2 volumes. Cambridge, MA: MIT Press.

Kueppers, Bernd-Olaf. 1987. "On the Prior Probability of the Existence of Life." In *The Probabilistic Revolution*, edited by Kreuger et al., vol. 2, Chapter 13. Cambridge, MA: MIT Press.

Lagueux, O. 2003. "Geoffroy's Giraffe: The Hagiography of a Charismatic Mammal." *Journal of the History of Biology* no. 36 (2):225–47.

Lamarck, J. B. 1801. *Systeme Des Animaux Sans Vertebres*. Paris: Deterville.

——. 1984. *Zoological Philosophy: An Exposition with Regard to the Natural History of Animals*. Chicago: University of Chicago Press.

Lande, R., Steiner Engen, and Berndt-Erik Saether. 2003. *Stochastic Population Dynamics in Ecology and Conservation: An Introduction*. Oxford: Oxford University Press.

Lang, H. S. 1983. "Aristotle and Darwin: The Problem of Species." *International Philosophical Quarterly* no. 23:141–53.
Laufer, B. 1928. "The Giraffe in History and Art." *Field Museum of Natural History*, 1–100. Chicago: Unknown Binding.
Le Guyader, Herve. 1998. *Etienne Geoffroy Saint-Hilaire 1772–1844: A Visionary Naturalist* (trans. Marjorie Grene). Chicago: University of Chicago Press.
Lennox, J. *Darwinism* 2004. Available from http://plato.stanford.edu/archives/fall2004/entries/darwinism/.
——. 1993. "Darwin Was a Teleologist." *Biology and Philosophy* 8: 409–21.
——. 2001. *Aristotle's Philosophy of Biology: Studies in the Origins of Life Science, Cambridge Studies in Philosophy and Biology*. Cambridge: Cambridge University Press.
Leonormand, T., Denis Rose, and Francois Roussert. 2009. "Stochasticity in Evolution." *Trends in Ecology and Evolution* 24: 157–65.
Lewens, T., D. Hull, and M. Ruse, eds. 1998. *The Philosophy of Biology*. Cambridge, MA: Cambridge University Press.
Lingwood, P. F. 1984. "The Duties of Natural History." *Biology Curators' Group Newsletter* no. 3 (9):531–33.
Løvtrup, S. 1987. *Darwinism: The Refutation of a Myth*. New York: Croom Helm in association with Methuen.
Lyell, C., and J. A. Secord. 1997. *Principles of Geology*. Penguin Classics. London: Penguin Books.
Lyell, Charles. 1830–3, first edition; 1837 5th edition. *Principles of Geology*. London: Murray.
——. 1863. *The Geological Evidences of the Antiquity of Man, with Remarks on the Origin of Species by Variation*. London: Murray.
Magner, L. N. 1994. *A History of the Life Sciences*. 2nd ed. New York: Marcel Dekker.
Manier, E. 1978. *The Young Darwin and His Cultural Circle: A Study of Influences Which Helped Shape the Language and Logic of the First Drafts of the Theory of Natural Selection*. Boston: D. Reidel Pub. Co.
Mayr, E. 1963. *Animal Species and Evolution*. Cambridge, MA: Belknap Press of Harvard University Press.
——. 1982. *The Growth of Biological Thought: Diversity, Evolution, and Inheritance*. Cambridge, MA: Belknap Press of Harvard University Press.
——. 1988. *Toward a New Philosophy of Biology: Observations of an Evolutionist*. Cambridge, MA: Belknap Press of Harvard University Press.
——. 1991. *One Long Argument: Charles Darwin and the Genesis of Modern Evolutionary Thought*. Cambridge: Harvard University Press.
Millstein, R. 2000. "Chance and Macroevolution." *Philosophy of Science* 67: 603–24.
——. 2006. "Discussion of 'Four Case Studies on Chance in Evolution: Philosophical Themes and Questions." *Philosophy of Science* 73:678–87.
——. 2011. "Chance and Causes in Evolutionary Biology: How Many Chances Become One Chance." *Philosophy of Science* 78:425–44.
Mivart, S. G. J. 1871. *On the Genesis of Species*. New York: D. Appleton and Company.
——. 1876. *Lessons from Nature, as Manifested in Mind and Matter*. London: J. Murray.
Moore, J. R. 1979. *The Post-Darwinian Controversies: A Study of the Protestant Struggle to Come to Terms with Darwin in Great Britain and America, 1870–1900*. Cambridge: Cambridge University Press.

——. 1985. "Darwin of Down: The Evolutionist as Squarson-Naturalist." In *The Darwinian Heritage*, edited by D. Kohn, 435–81. Princeton, NJ: Princeton University Press.

Nelson, G. J., and N. Platnick. 1981. *Systematics and Biogeography: Cladistics and Vicariance*. New York: Columbia University Press.

Nilsson, L. Anders. 1998. "Deep Flowers for Deep Tongues." *Trends in Ecology and Evolution*, 13, 259–60.

Nordenskiöld, N. E., and L. B. Eyre. 1929. *The History of Biology: A Survey*. London: Kegan Paul, Trench, Trubner & Co.

Ospovat, D. 1980. "God and Natural Selection: the Darwinian Idea of Design." *Journal of the History of Biology* 13: 169–94.

——. 1981. *The Development of Darwin's Theory: Natural History, Natural Theology, and Natural Selection, 1838–1859*. Cambridge: Cambridge University Press.

Outram, D. 1986. "Uncertain Legislator: Georges Cuvier's Laws of Nature in their Intellectual Context." *Journal of the History of Biology* 19: 323–68.

Owen, R. 1859. *On the Classification and Geographical Distribution of the Mammalia*. London: Parker and Son.

Packard, A. S. 1901. *Lamarck, the Founder of Evolution: His Life and Work*. New York and London: Longmans and Green.

Paley, W., J. Paxton, and J. Ware. 1841. *Natural Theology, or, Evidences of the Existence and Attributes of the Deity, Collected from the Appearances of Nature*. Boston: Gould, Kendall and Lincoln.

Peckham, M. 2006. *The Origin of Species: A Variorum Text*. Philadelphia: University of Pennsylvania Press.

Pellegrin, P., and A. Preus. 1986. *Aristotle's Classification of Animals: Biology and the Conceptual Unity of the Aristotelian Corpus*. Berkeley: University of California Press.

Pigliucci, M., and G.B. Müller, eds., 2010. *Evolution: The extended synthesis*. Cambridge, MA: MIT Press, xv–495pp.

Pincher, C. 1949. "Evolution of the Giraffe." *Nature* no. 164:29–30.

Pleins, J. David. 2013. *The Evolution of God: Charles Darwin and the Naturalness of Religion*. London: Bloomsbury Academic.

Quammen, D. 1997. *The Song of the Dodo: Island Biogeography in an Age of Extinctions*. New York: Simon & Schuster.

Raby, P. 2001. *Alfred Russel Wallace: A Life*. London: Chatto & Windus.

Radnitsky, G., and W. W. Barlow, eds. 1993. *Evolutionary Epistemology, Rationality, and the Sociology of Knowledge*. New York. Open Court.

Richards, R. A. 1997. "Darwin and the Inefficacy of Artificial Selection." *Studies in History and Philosophy of Science*, 28(1):75–97.

Richards, R. J. 1989. *Darwin and the Emergence of Evolutionary Theories of Mind and Behavior*. Chicago: University of Chicago Press.

——. Richards, R. J. 1992. *The Meaning of Evolution: The Morphological Construction and Ideological Reconstruction of Darwin's Theory, Science and Its Conceptual Foundations*. Chicago: Chicago University of Chicago Press.

Richardson, R. 2006. "Chance and the Patterns of Drift." *Philosophy of Science* 73:642–54.

Ridley, M. 1982. "Coadaptation and the Inadequacy of Natural Selection." *The British Journal for the History of Science* 15:45–68.

Romanes, G. 1897. *Darwin and after Darwin*. 3 vols. 2nd ed. Chicago: Open Court.

Rosenberg, A. 1985. *The Structure of Biological Science*. Cambridge: Cambridge University Press.

——. 2003. "Darwinism in Moral Philosophy and Social Theory," in *The Cambridge Companion to Darwin*, edited by J. Hodge; G. Radick, 310–32. Cambridge: Cambridge University Press.

Rudwick, M.J.S. 1970. "The Strategy of Lyell's *Principles of Geology*." *Isis* 61: 4–33.

——. 1982. "Charles Darwin in London: The Integration of Public and Private Science." *Isis* 73: 186–206.

Ruse, M. 1975. "Darwin's Debt to Philosophy: An Examination of the Influence of Philosophical Ideas on John F. W. Herschel and William Whewell on the Development of Charles Darwin's Theory of Evolution." *Studies in History and Philosophy of Science* (6):159–81.

——. 1978. "Darwin and Herschel." *Studies in History and Philosophy of Science* (9):323–31.

——. 1979. *The Darwinian Revolution*. Chicago: University of Chicago Press.

——. 1995. "Evolutionary Ethics: A Phoenix Arisen." In *Issues in Evolutionary Ethics*, edited by Paul Thompson, 245–247. Albany: State University of New York.

——. 2001. *Can a Darwinian Be a Christian? The Relationship between Science and Religion*. Cambridge: University of Cambridge Press.

——. 2003. *Darwin and Design: Does Evolution Have a Purpose?* Cambridge, MA: Harvard University Press.

Ruse, M., and R. J. Richards. 2008. *The Cambridge Companion to the "Origin of Species."* Cambridge: Cambridge University Press.

Schwartz, J. S. 1978. "The Genesis of Natural Selection—1838: Some Further Insights." *BioScience* no. 28 (5):321–26.

——. 1983. "Ideological and Intellectual Factors in the Genesis of the Theory of Natural Selection." In *Ideologie Et Al Revolution Darwiniere*, edited by Conroy. Paris: Vrin.

——. 1990. "Darwin, Wallace, and Huxley, and 'Vestiges of the Natural History of Creation.'" *Journal of the History of Biology* no. 23 (1):127–53.

Schweber, S. S. 1977. "The Origin of the 'Origin' Revisited." *Journal of the History of Biology* no. 10 (2):229–316.

——. 1978. "The Genesis of Natural Selection—1838: Some Further Insights." *BioScience* no. 28 (5):321–26.

——. 1979. "The Young Darwin." *Journal of the History of Biology* no. 12 (1):175–92.

——. 1980. "Darwin and the Political Economists: Divergence of Character." *Journal of the History of Biology* no. 13 (2):195–289.

——. 1994. "Darwin and the Agronomists: An Influence of Political Economy on Scientific Thought." *Boston Studies in the Philosophy of Science* no. 150:305–16.

Shanahan, T. 1991. "Chance as an Explanatory Factor in Evolutionary Biology." *History and Philosophy of the Life Sciences* 13: 249–269.

Shapin, S., and B. Barnes. 1979. "Darwinism and Social Darwinism: Purity and History." In *Natural Order: Historical Studies of Scientific Culture*, edited by B. Barnes and S. Shapin, 125–42. London: Sage Publications.

Sheets-Pyenson, S. 1981. "Darwin's Data: His Reading of Natural History Journals, 1837–1842." *Journal of the History of Biology* no. 14 (2):231–48.

Shepperson, G. 1961. "The Intellectual Background of Charles Darwin's Student Years at Edinburg." In *Darwinism and the Study of Society: A Centenary Symposium*, edited by M. Banton, 17–35. London: Tavistock Publications.

Singer, C. J. 1959. *A History of Biology to about the Year 1900: A General Introduction to the Study of Living Things*. 3d and rev. ed, The Life of Science Library. London: Abelard–Schuman.

Sloan, P. R. 1986. "Darwin, Vital Matter, and the Transformism of Species." *Journal of the History of Biology* no. 19 (3):369–445.

Smith, S. 1960. "The Origin of 'the Origin': As Discerned from Charles Darwin's Notebooks and His Annotations in the Books He Read between 1837 and 1842." *British Association for the Advancement of Science* no. 64:391–402.

Smith, S., and R. C. Stauffer. 1982. "New Light on the Foundations of the Origin of Species: A Reconstruction of the Archival Record." *Journal of the History of Biology* no. 15 (3):419–42.

Sober, E.. ed. 1983. *Conceptual Issues in Evolutionary Biology: An Anthology*. Cambridge, MA: MIT Press.

——. 1993. *Philosophy of Biology*. Boulder, CO: Westview Press (in UK: Oxford University Press).

——. 2000. "Psychological Egoism" in *The Blackwell Guide to Ethical Theory*, edited by Hugh Lafollette, 129–148. London: Wiley-Blackwell Publishing Ltd.

Southward, A. J. 1983. "A New Look at Variation in Darwin's Species of Acorn Barnacles." *Biological Journal of the Linnean Society* no. 20 (1):59–72.

Spencer, H. 1866. *The Principles of Biology*. 2 vols. New York: D. Appleton and Company.

Stevens, P. F. 1995. "George Bentham and the Darwin/Wallace Papers of 1858: More Myths Surrounding the Origin and Acceptance of Evolutionary Ideas." *Linnean* no. 11:14–16.

Stone, I., and J. Stone. 1980. *The Origin: A Biographical Novel of Charles Darwin*. 1st ed. New York: Doubleday.

Stott, R. 2012. *Darwin's Ghosts*. New York: Spiegel and Grau.

Sulloway, F. J. 1979. "Geographic Isolation in Darwin's Thinking: The Vicissitudes of a Crucial Idea." *Studies in the History of Biology* no. 3: 23–65.

——. 1982. "Darwin's Conversion: The Beagle Voyage and Its Aftermath." *Journal of the History of Biology* 15 (3):325–96.

——. 1982. "Darwin and His Finches: The Evolution of a Legend." *Journal of the History of Biology* no. 15 (1):1–53.

——. 1985. "Darwin's Early Intellectual Development: An Overview of the Beagle Voyage (1831–1836)." In *The Darwinian Heritage*, edited by D. Kohn, 121–54. Princeton, NJ: Princeton University Press.

Thagard, Paul. 1977. "Darwin and Whewell." *Studies in History and Philosophy of Science* (8):353–6.

Thomson, K. S. 2005. *Before Darwin: Reconciling God and Nature*. New Haven, CT: Yale University Press.

Turbayne, C. 1970. *The Myth of Metaphor*. Columbia: University of South Carolina Press.

Turner, John R. 1987. "Random Genetic Drift, R. A. Fisher, and the Oxford School of Ecological Genetics." In *The Probabilistic Revolution*, edited by Kreuger et al., vol. 2, ch. 12. Cambridge, MA: MIT Press.

Vorzimmer, P. J. 1964. *The Development of Darwin's Evolutionary Thought after 1859*, Cambridge: Cambridge University Press.

——. 1969. "Darwin, Malthus, and the Theory of Natural Selection." *Journal of the History of Ideas* no. 30:527–42.

Vorzimmer, P. J., and C. Darwin. 1969. "Darwin's "Questions about the Breeding of Animals" (1839)." *Journal of the History of Biology* no. 2 (1):269–81.

——. 1969. "Darwin's 'Lamarckism' and the 'Flat-Fish Controversy': 1863–1871." *Lychnos* 121–70.

——. 1970. *Charles Darwin: The Years of Controversy; the Origin of Species and Its Critics, 1859–1882*. Philadelphia: Temple University Press.

——. 1977. "The Darwin Reading Notebooks (1838–1860)." *Journal of the History of Biology* no. 10 (1):107–53.

Wallace, A. R. 1903. "My Relations with Darwin in Reference to the Theory of Natural Selection." *Black and White*, January 1903, 78–79.

Wells, K. D. 1973. "The Historical Context of Natural Selection: The Case of Patrick Matthew." *Journal of the History of Biology* no. 6 (2):225–58.

Whewell, William. 1833. *Astronomy and General Physics Considered with Reference to Natural Theology* (Bridgewater Treatise), London: William Pickering.

——. 1840. *Philosophy of the Inductive Sciences*. London: John Parker.

White, M., and J. Gribbin. 1995. *Darwin: A Life in Science*. New York: Dutton.

Wilkins, J. 2003. The Talk Origins Archive: Exploring the Creation/Evolution Controversy. *Darwin's Precursors and Influences*, http://www.talkorigins.org/.

Wolters, G., James G. Lennox, and Peter McLaughlin. 1995. *Concepts, Theories, and Rationality in the Biological Sciences: The Second Pittsburgh-Konstanz Colloquium in the Philosophy of Science*. Pittsburgh, PA: University of Pittsburgh Press.

Wright, C. 1871. "The Genesis of Species." *North American Review* no. 113:63–103.

Wright, Chauncy. (1871), "Review of Mivart 1871" in *N. American Review* no. 113, 86–9.

Wyhe, J. V. 2002. Charles Darwin's Earliest Doubts Concerning the Immutability of Species. *The Complete Work of Charles Darwin Online*, http://darwin-online.org.uk/.

Young, R. M. 1969. "Malthus and the Evolutionists: The Common Context of Biological and Social Theory." *Past and Present* no. 43 (1):109–45.

——. 1985. *Darwin's Metaphor*. Cambridge: Cambridge University Press.

GENERAL INDEX

INDEX NOMINUM